TRIGONOMETRY

Notes, Problems and Exercises

TRIGONOMETRY

Notes, Problems and Exercises

Roger Delbourgo

Formerly at Middlesex University, UK

World Scientific

NEW JERSEY · LONDON · SINGAPORE · BEIJING · SHANGHAI · HONG KONG · TAIPEI · CHENNAI · TOKYO

Published by

World Scientific Publishing Co. Pte. Ltd.

5 Toh Tuck Link, Singapore 596224

USA office: 27 Warren Street, Suite 401-402, Hackensack, NJ 07601

UK office: 57 Shelton Street, Covent Garden, London WC2H 9HE

Library of Congress Cataloging-in-Publication Data
Names: Delbourgo, Roger.
Title: Trigonometry : notes, problems, and exercises / by Roger Delbourgo
    (Middlesex University, UK).
Description: New Jersey : World Scientific, 2017. |
    Includes bibliographical references and index.
Identifiers: LCCN 2016052208| ISBN 9789813207103 (hardcover : alk. paper) |
    ISBN 9789813203112 (pbk : alk. paper)
Subjects: LCSH: Trigonometry--Textbooks.
Classification: LCC QA531 .D37 2017 | DDC 516.24--dc23
LC record available at https://lccn.loc.gov/2016052208

British Library Cataloguing-in-Publication Data
A catalogue record for this book is available from the British Library.

Printed in Singapore

Preface

Trigonometry is predominantly about measurements on triangles, and has for its original aim "the solution of triangles". The dictionary defines it as the branch of Mathematics concerned with relations between the sides and angles of a triangle. The subject evolved from the investigations carried out by the Middle Eastern astronomer Hipparchus, c. 140 BC. Three hundred years later, the next major contributor was Ptolemy of Alexandria, publishing his Almagest, c. 140 AD. Due to the researches of many workers in this field, the subject today has become quite vast and it now occupies a central position in Science, with applications in most areas of life.

The primary purpose of the book is to provide, in a series of short Chapters, a coherent account of Trigonometry that would benefit students following a science-based career. Our survey includes many standard results which are intricately connected. We thus hope to reveal the beauty and elegance of the subject — our secondary purpose. The book covers the elements of Plane Trigonometry but a digression is made in the Appendix, where some useful aspects of practical Spherical Trigonometry are introduced.

Geometry lies at the core of this book. Changes of teaching emphasis over the years have resulted in the virtual disappearance of geometry from the curriculum and this is a quite regrettable development. It is to remedy this state of affairs that we took the conscious decision of supplying some of the long-neglected geometry. Most of the finest trigonometric facts are reviewed and proved again. We mention some of the famous contributors, namely: Pythagoras, Apollonius, Hero, Ptolemy, Euler, Napier, Brahmagupta, Feuerbach, Morley, Napoleon, Fermat, ... In any trigonometric presentation we should not ignore the remarkable theorems that have come to us from the past. Unfortunately in these few pages the entire story cannot be told.

To give substance to the bookwork, many fully-worked examples are supplied. The reasons for including many are various: foremost among them is the wish to indicate the kind of problems that might be posed. A second reason is to show how progress through a solution is made; studying the sequence of steps involved makes one realise why they need to be followed. A third reason is to explore ramifications of the theorems. We suggest that a problem should be tried first before a comparison is made with the given solution. Valuable experience in the art of problem-solving may be gained that way and short exercises have been set, extending some of the problems. For further practice, especially in preparing for an examination, it is naturally best to consult past question papers.

The book is organised into short Chapters, each treating one specific theme and are a few pages long; they should preferably be read in order, so as to preserve some continuity in the development. Readers needing a shorter course may omit the following Chapters: 18, 27–29, 31, 32, if they wish, as well as the Appendix on Spherical Trigonometry.

The basic results are established early on. The sine and cosine rules lead to the factorisation and addition formulae on which much else depends. Ptolemy's theorem is proved, identities and equations are considered, and we deal also with elimination questions, series summation and Euler's product. Morley's amazing theorem is proved trigonometrically. We study the location of special points in a triangle and obtain the angle and distance formulae related to the points. The nine-point circle theorem is treated, so are Napoleon's circles. Thus we discover inter-relations between the points O, I, H, N, G, F.

There is a chapter on cyclic quadrilaterals. Completion of a proof is indicated by a □.

Regarding angular measurements, a start is made in Chapter 2, where the degrees to radians conversion is made, but radians are not used to any extent until very much later when it is absolutely vital to do so. Until then we may think of angles as being reckoned in degrees. The full advantage of working in radians will be seen when more theoretical and practical problems are presented.

In the Appendix, following a suggestion by my brother Robert, I have introduced some Spherical Trigonometry. We prove here the spherical counterparts of the sine and cosine rules, applying them to solve interesting problems related to the Earth.

The entire contents are quite elementary, in that advanced methods have not been used. On one occasion only (in the Appendix) we have used the dot product from vector algebra. The problems and exercises in the main are straightforward applications of the theorems.

In conclusion, I should like to thank Prakash Mehta, MEng, for the enormous help he afforded me in computerising all of my work. Without him these notes would not have seen the light of day. Also, I must thank my ex-colleague and friend, Dr Ramon Prasad, for his valued support throughout this undertaking.

Roger Delbourgo
Chiswick, London W4
2016

P.S.

Following a recommendation from WSPC and the reviewers, I have added appendix B and C to the book, wherein are contained a selection of extra examples/problems with their solutions. This extends the book by almost 30% to mini-textbook size, and serves to provide further practice for the more enthusiastic students.

Contents

Chapter 0

Introduction. A Review of Some Geometrical Ideas

0.1 Assumptions

Before beginning any serious study of Trigonometry, the reader should appreciate that a solid grounding in Elementary Mathematics is essential. An adequate amount of Geometry should have been studied at school level. In this regard, knowledge of all of the following topics is necessary before any progress is possible.

Properties of concurrent and parallel lines; similarity of triangles and congruence tests; parallelograms, proportion theorems, areas; the circle theorems, angle and distance relationships in respect of chords, tangents and secants; some elementary facts about vectors; the theorem of Pythagoras.

The last topic will mark the start of our own development (Chapter 1).

0.2 Notation

We deal mostly with triangles. ABC denotes a triangle with vertices A, B, C; $\triangle$, standing alone will also denote its area. The angles of this triangle, namely $B\hat{A}C$, $C\hat{B}A$, $A\hat{C}B$, usually written $\hat{A}$, $\hat{B}$, $\hat{C}$ will be abbreviated to A, B, C, when no confusion with the vertices is possible.

It is a tradition dating from Euler's period (1707–1783) that the sides BC, CA, AB are represented as a, b, c (respectively). In formulae, we frequently use the semi-perimeter of the $\triangle ABC : s = \frac{1}{2}(a + b + c)$. Yet other special letters will appear in due course, e.g. R, r. All of them will be defined later.

General angles will bear Greek letters: θ (theta), ϕ (phi), α (alpha), $\cdots$, measured in degrees or radians.

0.3   Special points

A number of special points associated with a $\triangle ABC$ will be considered soon. Some may already be known. Prominent amongst these are:

(i) O, the centre of the circumscribing circle where the perpendicular bisectors of the sides all meet. (Chapter 6)

(ii) G, the point of intersection of the three medians. (Chapter 5)

(iii) I, the centre of the circle touching all three sides internally; it is where the angle bisectors meet. (Chapter 25)

(iv) H, the point of intersection of the altitudes drawn from the vertices to the opposite sides. (Chapter 26)

(v) N, the centre of the pedal (or 'orthic') triangle. (Chapter 30)

(vi) F, the Fermat point. (Chapter 32)

It is a fact that O, G, N, H are collinear (i.e. lie on a straight line). This will be proved later. As will be seen, there are many other interesting results to be discovered.

0.4   Problems / exercises, reviewing some geometrical ideas

We have selected a few problems from geometry below, and given full so- lutions in order to highlight some important ideas. Of course, not all areas of elementary geometry can be covered, but the selection should help set the scene for the trigonometry that is to follow.

1) Refer to Fig. 0.4.1.
$KL \parallel BC$, and $AK = 3\,\mathrm{cm}$, $KL = 4\,\mathrm{cm}$, $LC = 5\,\mathrm{cm}$, $BC = 10\,\mathrm{cm}$.

Find $\dfrac{\text{area}\,KLCB}{\text{area}\,\triangle ABC}$.

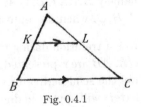

Fig. 0.4.1

Solution. Since $KL \parallel BC, \triangle AKL$ is similar to $\triangle ABC$, so

$$\frac{AK}{AB} = \frac{AL}{AC} = \frac{KL}{BC} = \frac{4}{10} = 0.4.$$

In consequence, $\dfrac{\text{area} \triangle AKL}{\text{area} \triangle ABC} = 0.4^2 = 0.16$, and the required ratio is

$$1 - 0.16 = 0.84.$$

Exercise. Using the ratios in the solution, show that $KB = 4\frac{1}{2}$ cm, and $AL = 3\frac{1}{3}$ cm. ◇

2) Refer to Fig. 0.4.2.
AX bisects $\hat{A}$ of the triangle. $AB = 6\,\text{cm}$, $AC = 4\,\text{cm}$, $BC = 5\,\text{cm}$. Find XC.

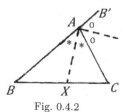

Fig. 0.4.2

Solution. Since AX bisects $\hat{A}$, $\frac{BX}{XC} = \frac{AB}{AC}$, hence $\frac{5-XC}{XC} = \frac{6}{4}$, $10XC = 20$, $XC = 2\,\text{cm}$.

Exercise. The external bisector of $\hat{A}$, shown in the above, meets BC produced at Y. From the fact that $\frac{BY}{CY} = \frac{AB}{AC}$, find CY and hence XY.
Ans: $CY = 10\,\text{cm}$, $XY = 12\,\text{cm}$. ◇

3) Refer to Fig. 0.4.3.
$ADOD'$ is a line through the centre O of a circle.
ACB is a secant; AT is tangential to the circle.
Given $BC = 5\,\text{cm}$, $CA = 4\,\text{cm}$, $AD = 3\,\text{cm}$. Find AT and DD'.

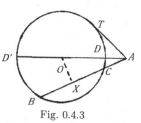

Fig. 0.4.3

Solution. It is known that $AT^2 = AC \cdot AB = AD \cdot AD'$.

From the first equality, $AT^2 = 4(4 + 5) = 36$, $\therefore AT = 6\,\text{cm}$. From the second equality, therefore, $36 = 3(3 + DD')$,

$\therefore DD' = 9\,\text{cm}$.

Exercise. In the above, if $OX \perp BC$, find OX from $\triangle OXC$ (or $\triangle OXB$). Ans: $OX = \sqrt{14}$ cm.   ◇

4) Refer to Fig. 0.4.4.
Point P is taken on the circumference of $\triangle ABC$.
L, M, N are the feet of the perpendiculars from
P to the sides of the triangle.
Show L, M, N are collinear points.
The line NLM is Simpson's Line for P.

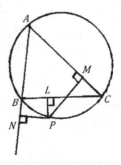

Solution. Because of the right angles at L, M, N
it is evident that quadrilaterals $PLBN, PLMC$
are both cyclic.

Fig. 0.4.4

$PBAC$ is obviously cyclic. From properties of these cyclic figures, we find

$$P\hat{L}N = P\hat{B}N$$
$$= P\hat{C}A$$
$$= 180° - P\hat{L}M.$$

Thus $P\hat{L}N + P\hat{L}M = 180°$, implying NLM is a straight line.

5) Refer to Fig. 0.4.5.
$OABC$ is a parallelogram. The positions of A, C relative to O are given by
the vectors $\overline{OA} = \underline{a}$, $\overline{OC} = \underline{c}$.
X is taken on CB with $CX : XB = 2 :$
1; Y is the midpoint of OX, and AY
cuts OC in Z.

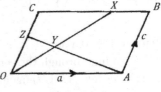

Fig. 0.4.5

Show that $\overline{OY} = \frac{1}{3}\underline{a} + \frac{1}{2}\underline{c}$ and that $OZ : OC = 3 : 1$.

Solution. Because $OABC$ is a parallelogram,

$\overline{CB} = \overline{OA} = \underline{a}$ and $\overline{AB} = \overline{OC} = \underline{c}$.

Since X is such that $CX : XB = 2 : 1$, $\overline{CX} = \frac{2}{3}\underline{a}$.

Then by the addition rules for vectors $\overline{OX} = \overline{OC} + \overline{CX} = \underline{c} + \frac{2}{3}\underline{a}$.

Y, being the midpoint of OX, we have $\overline{OY} = \frac{1}{2}\overline{OX} = \frac{1}{3}\underline{a} + \frac{1}{2}\underline{c}$.

This makes $\overline{AY} = \overline{AO} + \overline{OY} = -\underline{a} + (\frac{1}{3}\underline{a} + \frac{1}{2}\underline{c}) = -\frac{2}{3}\underline{a} + \frac{1}{2}\underline{c}$.

$\overline{AZ}$ is a multiple of $\overline{AY}$:

$$\overline{AZ} = m \cdot \overline{AY} = m(-\frac{2}{3}\underline{a} + \frac{1}{2}\underline{c}),$$

so making $\overline{OZ} = \overline{OA} + \overline{AZ} = \underline{a} + m(-\frac{2}{3}\underline{a} + \frac{1}{2}\underline{c}) = (1 - \frac{2}{3}m)\underline{a} + \frac{1}{2}m\underline{c}$.

But Z is on OC, so $1 - \frac{2}{3}m = 0, m = \frac{3}{2}$.

Therefore $\overline{OZ} = \frac{1}{2}m\underline{c} = \frac{3}{4}\underline{c}$.

Thus $OZ : ZC = 3 : 1$.

The theorems enabling most of the solutions to be effected in the above problems can be found as numbered propositions in Euclid's books.

Exercise. If in the above CY meets OA in W, find $OW : WA$.

<u>Ans:</u> $2 : 1$.

◇

Chapter 1

Pythagoras' Theorem

1.1 Historical

Pythagoras was a Greek philosopher interested in the mystical properties of numbers. The Pythagoreans are credited with a proper proof of the famous theorem, but it is believed that the Babylonians knew of the result centuries earlier. The classical proof is in the Elements of Euclid (c. 300 BC). Euclid founded the Alexandrian School of Mathematics but his multivolume work was printed only in 1482 AD. Well over 350 proofs exist today. We give, below, the classical proof and one other based on similar triangles.

1.2 Theorem [Pythagoras of Samos, 585–501 BC]

Theorem 1.1 (Theorem of Pythagoras). *In a right-angled triangle the square of the hypotenuse equals the sum of the squares of the other two sides. E.g., if in $\triangle ABC$, $A = 90°$, then (in the standard notation referred to in 0.2.),*

$$a^2 = b^2 + c^2,$$

a being the hypotenuse; b, c the other sides (or legs).

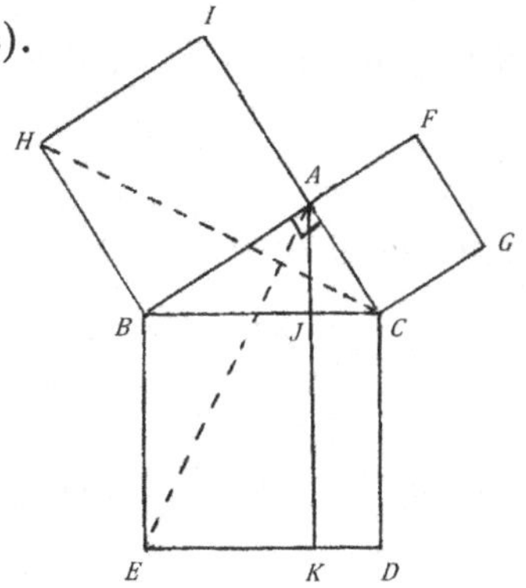

Fig. 1.2.1

Classical proof. Refer to Fig. 1.2.1.
$\triangle ABC$ has $A = 90°$. Squares $BCDE$, $CAFG$, $ABHI$ are drawn outside

7

the triangle. We have to prove

(†) area sq. $BCDE$ = area sq. $CAFG$ + area sq. $ABHI$.

Join A to E, and C to H and draw $AJK \perp BC$.

Now $\triangle ABE$ is congruent to $\triangle HBC$; this is because the former can be brought into coincidence with the latter by a $90°$ anti-clockwise rotation about the point B.

Area $\triangle ABE = \frac{1}{2}BE \cdot BJ = \frac{1}{2}$ area $BJKE$;

and area $\triangle HBC = \frac{1}{2}HB \cdot AB = \frac{1}{2}$ area $ABHI$.

$\therefore$ area $BJKE$ = area $ABHI = c^2$.

Similarly, by joining B to G, and A to D, we prove area $CJKD$ = area $CAFG = b^2$.

Adding the last pair of results gives (†):

$$\text{area } BCDE = a^2 = c^2 + b^2.$$

$\square$

A second proof. Refer to Fig. 1.2.2.

Draw $AX \perp BC$.

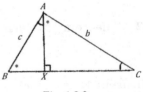

The construction makes triangles ABC, XBA, XAC similar to one another.

Fig. 1.2.2

$\triangle ABC$ has hypotenuse $BC = a$;

$\triangle XBA$ has hypotenuse $AB = c$;

$\triangle XAC$ has hypotenuse $AC = b$.

We therefore have

$$\frac{\text{area } \triangle ABC}{a^2} = \frac{\text{area } \triangle XBA}{c^2} = \frac{\text{area } \triangle XAC}{b^2} = k, \text{ say.}$$

But area $\triangle ABC$ = area $\triangle XBA$ + area $\triangle XAC$, so $ka^2 = kc^2 + kb^2, a^2 = b^2 + c^2$.

$\square$

The converse of Pythagoras' theorem holds; i.e., if between the sides of a triangle a relation of the type $a^2 = b^2 + c^2$ exists, then the triangle must be right-angled, with the right-angle opposite the longest side.

1.3 Problems / exercises

1) A ladder 8.5 m long rests with its base 4.0 m away from a vertical wall of a house to reach a window. How high does the ladder reach up the wall?

Solution. If the height is h metres then by Theorem 1.1,

$8.5^2 = 4.0^2 + h^2$,

$h = \sqrt{8.5^2 - 4.0^2} = 7.5$ metres, exactly.

2) Check that the triangle with sides 40 cm, 55 cm, 73 cm is right-angled

Solution. $48^2 + 55^2 = 2304 + 3025 = 5329$, and $73^2 = 5329$.

So the triangle is right-angled.

Exercise. Which of the following triangles are right-angled? The sides are given as

i) 9, 40, 41; ii) 8, 15, 17; iii) 11, 13, 17.

<u>Ans:</u> The first two only. ◇

Chapter 2

Degrees versus Radians

Remark 2.1. We will normally be measuring all our angles in degrees. This will be the case from Chapter 3 through to Chapter 32 inclusive. Although a straightforward conversion is all that is involved, the necessity to use radians, as such will only become apparent much later, as the discussions become more theoretical. In this chapter we lay only the foundations and show at the same time that radian measure can have practical uses. But for the time being we shall express (from Chapter 3 onwards) all results in terms of angles expressed in degrees.

2.1 An easy conversion

Refer to Fig. 2.1.1.

On a circle, radius r, centre O, two points A, P are marked. Let $A\hat{O}P = \theta°$, length of arc $AP = s$, and area of sector $AOP = S$. By simple proportion, $\frac{\theta}{360} = \frac{s}{2\pi r} = \frac{S}{\pi r^2}$.

This yields

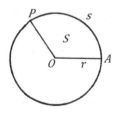

Fig. 2.1.1

$(*)\quad s = \pi r\left(\frac{\theta}{180}\right),\quad S = \frac{1}{2}\pi r^2\left(\frac{\theta}{180}\right).$

If $s = r$, then $\theta = \frac{180}{\pi}(\simeq 57.3)$. This value is 1 RADIAN. Thus we set $180° = \pi$ rads (i.e. $3.14159...$ rads) or inversely $1\,\mathrm{rad} = \frac{180}{\pi}$ degrees. If $\theta° = \alpha$ rads, formulae $(*)$ may therefore be replaced by

$(**)\quad s = r\alpha,\quad S = \frac{1}{2}r^2\alpha.$

11

As indicated before, theoretical reasons for converting to radians will emerge later.

For future reference it is wise to commit to memory the results

$30° = \frac{\pi}{6}$ rads, $45° = \frac{\pi}{4}$ rads, $60° = \frac{\pi}{3}$ rads.

2.2   Problems / exercises

1) Refer to Fig. 2.2.1.

A wire of total length L is bent to the shape of a sector. Express the enclosed area, S, in terms of L and the length of each straight part, r.

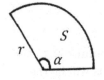

Deduce that the maximum value of $S = \frac{L^2}{16}$ when the sectorial angle $\alpha = 2$ rads.

Fig. 2.2.1

Solution. From $(**)$, $L = 2r + r\alpha$ and $S = \frac{1}{2}r^2\alpha$. So $\alpha = \frac{L}{r} - 2$, $S = \frac{1}{2}r^2(\frac{L}{r} - 2) = \frac{1}{2}Lr - r^2 = \frac{L^2}{16} - (r - \frac{L}{4})^2$, on completing the square in the quadratic.

Thus when $\alpha = 2$, $r = \frac{L}{4}$ and S achieves its maximum value of $\frac{L^2}{16}$.

2) Refer to Fig. 2.2.2.

A conical lampshade is made by folding the sectorial shape $OACB$, so OA, OB are joined together (as in (b)). If the sectorial angle, $\alpha = \frac{6\pi}{5}$ (rads), find the base radius R and height h of the cone in terms of the slant edge r of the cone.

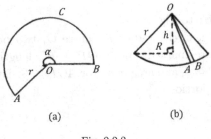

(a)   (b)

Fig. 2.2.2

Solution. Note $\alpha = \frac{6\pi}{5}$ rads $= 216°$ (exactly).

$r = OA = OB$, arc $ACB = r\alpha = \frac{6\pi r}{5}$ (from $(*)$) = circumference of conical base, after the folding operation.

∴ base radius of the shade is $R = \dfrac{\frac{6\pi r}{5}}{2\pi} = \dfrac{3r}{5}$.

By 1.2. the height of the shade is $h = \sqrt{r^2 - (\frac{3r}{5})^2} = \frac{4r}{5}$.

3) Refer to Fig. 2.2.3.

On a disc of radius r, centre O, a chord AB of length r is drawn. This chord divides the disc into minor and major segments of areas m, M (respectively). Find m and M.

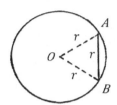

Fig. 2.2.3

Solution. The distance from O to AB = $\sqrt{r^2 - (\frac{r}{2})^2} = \frac{r\sqrt{3}}{2}$.

∴ area $\triangle OAB = \frac{1}{2}r\left(\frac{r\sqrt{3}}{2}\right) = \frac{r^2\sqrt{3}}{4}$.

By 2.1., area of minor sector $OAB = \frac{1}{2}r^2\frac{\pi}{3} = \frac{\pi r^2}{6}$.

∴ m = area sector OAB − area $\triangle OAB = \frac{\pi r^2}{6} - \frac{r^2\sqrt{3}}{4} = \frac{r^2}{12}(2\pi - 3\sqrt{3})$.

Then M = area of major segment = $\pi r^2 - m = \frac{r^2}{12}(10\pi + 3\sqrt{3})$.

Exercise. Show that $\frac{M}{m} \simeq 33.68$.

◇

Chapter 3

Cartesian and Polar Coordinates. The Sine and Cosine Ratios

3.1 Cartesians and polars

Refer to Fig. 3.1.1.

Cartesian coordinates [René Descartes, 1596–1650] are already familiar to all readers. Axes $x'Ox$, $y'Oy$ on a plane are chosen (generally and preferably at right angles to each other), and a scale of lengths established. The axes divide the plane into 4 numbered quadrants, as indicated. Points P_i $(i = 1, \cdots, 4)$ are selected, one in each of the quadrants. Their distances from the origin O are $OP_i = r_i$; $r_i > 0$. Assume no P_i is located at O.

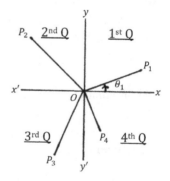

Fig. 3.1.1

$x\hat{O}P_i = \theta_i$ $(i = 1, \cdots, 4)$. Any point P has cartesian coordinates (x, y) and polar coordinates $[r, \theta]$.

Clearly, $r = \sqrt{x^2 + y^2}$.

The angles θ_i are measured anti-clockwise from Ox and are regarded as positive. But negative angles, measured clockwise from Ox may be considered, and we regard a negative angle φ_i as being equivalent (at least in so far as trigonometric ratios are concerned) to the positive angle $360° + \varphi_i$. So, e.g., $240°$ is equivalent to $-120°$; and $-330°$ is equivalent to $+30°$.

For example: $(-5,0) \equiv [5,-180°]$ and $(0,3) \equiv [3,90°] \equiv [3,-270°]$.

3.2   Definitions

Given (x,y) with corresponding polar coordinates $[r,\theta]$, we define

cosine of θ, $\cos\theta = \frac{x}{r}$ and sine of θ, $\sin\theta = \frac{y}{r}$ $(r>0)$.

These ratios clearly depend on the relative values of x,y,r and not on their actual magnitudes. Notice that if θ is acute, $\cos\theta$ and $\sin\theta$ are positive $(\because x,y>0)$ whereas if θ is obtuse, $\cos\theta$ is negative $(\because x<0)$ but $\sin\theta$ is positive $(\because y>0)$.

Exercise. a) Setting the calculator to "degree-mode" verify the following values: (to 4 d.p.)

$$\sin 10° \simeq 0.1736, \cos 35° \simeq 0.8192, \sin 100° \simeq 0.9848, \cos 170° \simeq -0.9848.$$

b) Setting the calculator to "radian-mode", find

$$\sin\frac{\pi}{10}, \cos\frac{5\pi}{9}, \sin\frac{7\pi}{6}, \cos\frac{3\pi}{8}.$$

<u>Ans:</u> Approximately $0.3090, -0.1736, -0.5, 0.3827$ respectively.   ◇

3.3   Theorem

Theorem 3.1. *For any given* θ,
  (i) $-1 \le \cos\theta \le 1, -1 \le \sin\theta \le 1$ *and* $(\cos\theta)^2 + (\sin\theta)^2 = 1$;
  (ii) $\cos(-\theta) = \cos\theta, \quad \sin(-\theta) = -\sin\theta$;
  (iii) $\cos(90° \mp \theta) = \pm\sin\theta, \quad \sin(90° \mp \theta) = \cos\theta$;
      $\cos(180° \mp \theta) = -\cos\theta, \quad \sin(180° \mp \theta) = \pm\sin\theta.$

Remark 3.1. The identity in part (i) is usually abbreviated to $\cos^2\theta + \sin^2\theta = 1$. It is not to be confused with $\cos(\cos\theta) + \sin(\sin\theta) = 1$; which is altogether different and wrong. (ii) shows that $\cos\theta$ is an even function, while $\sin\theta$ is an odd function. A function f defined on an interval $-\alpha \le \theta \le \alpha$ is even if $f(\theta) = f(-\theta)$ and is odd if $f(\theta) = -f(-\theta)$. In (iii), either upper signs are used or else lower signs. There are 8 results in all.

Proof. Take P with coordinates (x, y) or $[r, \theta]$.

(i) Since $|x| \leq r$, $|y| \leq r$, we have $-1 \leq \frac{x}{r} \leq 1$, $-1 \leq \frac{y}{r} \leq 1$, i.e. $-1 \leq \cos \theta \leq 1$, and $-1 \leq \sin \theta \leq 1$.

Further, $\cos^2 \theta + \sin^2 \theta = (\frac{x}{r})^2 + (\frac{y}{r})^2 = \frac{x^2 + y^2}{r^2} = 1$.

(ii), (iii): we prove a sample of these identities.

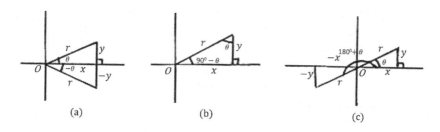

(a)   (b)   (c)

Fig. 3.3.1

Refer to Fig. 3.3.1.

In (a), $\cos(-\theta) = \frac{x}{r} = \cos \theta$, $\sin(-\theta) = \frac{-y}{r} = -\sin \theta$, proving (ii) fully.

In (b), $\cos(90° - \theta) = \frac{x}{r} = \sin \theta$.

In (c), $\sin(180° + \theta) = \frac{-y}{r} = -\sin \theta$. This proves the first and eighth identities in (iii). We have taken θ to be acute in the diagrams but this need not be so. ☐

Exercise. In a similar manner prove the remaining 6 identities in (iii). ◇

3.4   Problems / exercises

1) Refer to Fig. 3.4.1.

If θ is an obtuse angle such that $\sin\theta = \frac{7}{25}$, find $\cos\theta$.

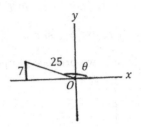

Solution. From part (i) in the theorem, $\cos^2\theta = 1 - \sin^2\theta$.

So, $\cos^2\theta = 1 - \left(\frac{7}{25}\right)^2 = \frac{576}{625}$.

Fig. 3.4.1

$\therefore \cos\theta = -\frac{\sqrt{576}}{\sqrt{625}} = -\frac{24}{25}$ exactly, which is negative because θ is obtuse.

2) Refer to Fig. 3.4.2.

A boat sails a distance 8 km from a marina O on a bearing of 45° (i.e. N 45° E) to a port of call, C. Subsequently it proceeds on a bearing of 35° (i.e. N 35° W) a further distance of 3 km to a final destination, D.

How far is D East of the marina?

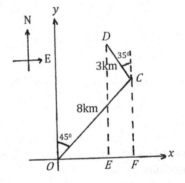

Fig. 3.4.2

Solution. An aerial view is shown. OE, OF are the projections of OD, OC in the Easterly direction. OE is the required distance East.

$OE = OF - EF = 8\sin 45° - 3\sin 35° = 3.94$ km.

Exercise. In the problem, how far is D North of O? Find also the direct distance OD.

<u>Ans:</u> 8.11 km due N, 9.02 km.   ◇

3) Refer to Fig. 3.4.3.

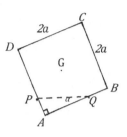

A uniform square tile $ABCD$, of side $2a$, rests vertically on horizontal supports P, Q a distance c apart, with AB in contact with Q and making an angle α with the horizontal. How high is the centre of gravity G of the tile above the level of the pegs?

Fig. 3.4.3

Solution. Point A lies at a distance $AQ \sin \alpha$ below the horizontal line PQ.

Now $AQ = c \cos \alpha$.

$\therefore$ distance of A below $PQ = c \cos \alpha \sin \alpha$.

G is at a distance above A of $a \cos \alpha + a \sin \alpha$.

Hence G is at a height above $PQ = a(\cos \alpha + \sin \alpha) - c \cos \alpha \sin \alpha$.

Exercise. In the problem, find GP.

Ans: $\sqrt{a^2 + (a - c \sin \alpha)^2}$.

◇

Chapter 4

The Cosine Rule

4.1 Introduction

Also known as the Law of Cosines, the cosine rule constitutes a possible first step towards "solving a triangle", i.e., finding unknown sides or angles in a triangle. It is the trigonometric version of the extensions to the Pythagoras theorem, and is due to François Viète. It is used in one of two ways: (a) to find the missing side of a triangle when the two sides and the included angle are known ("SAS"); (b) to find the angles of a triangle when all three sides are known ("SSS").

Frequently in order to completely solve a triangle, the cosine rule is complemented by the sine rule; this rule appears in Chapter 6. We wish to remind readers of the naming conventions for sides and angles in a triangle. (See Chapter 0.2.)

4.2 The cosine rule [François Viète, 1540–1603]

Theorem 4.1 (The Cosine Rule). *In any $\triangle ABC$,*

$$a^2 = b^2 + c^2 - 2bc \cos A.$$

Or, equivalently,

$$\cos A = \frac{b^2 + c^2 - a^2}{2bc}.$$

Proof. (We also have $b^2 = c^2 + a^2 - 2ca \cos B$, $c^2 = a^2 + b^2 - 2ab \cos C$, which are obtainable through a cyclic interchange of letters.)

Refer to Fig. 4.2.1.

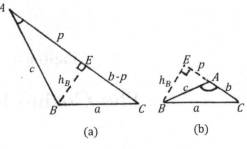

In (a), A is acute; in (b), A is obtuse.

Draw the height $h_B = BE$ from B to AC.

E is internal to AC in (a), but external to AC in (b).

Fig. 4.2.1

Let $p = AE$, the projected length of c onto AC.

Then $a^2 = \begin{cases} h_B^2 + (b-p)^2 & \text{in (a)}, \\ h_B^2 + (b+p)^2 & \text{in (b)}. \end{cases}$

But $h_B^2 = c^2 - p^2$ in either case, so

$$a^2 = \begin{cases} b^2 + c^2 - 2bp & \text{in (a)}, \\ b^2 + c^2 + 2bp & \text{in (b)}. \end{cases}$$

Now, by the definition of cosine θ in 3.2.

$\frac{p}{c} = \cos A$ in (a), and $\frac{p}{c} = -\cos A$ in (b).

In either event, therefore,

$$a^2 = b^2 + c^2 - 2bc \cos A.$$

Recasting gives $\cos A$ in terms of the sides. $\qquad\square$

4.3   Problems / exercises

1) Refer to Fig. 4.3.1.

$\triangle ABC$ has $AB = 5\,\text{cm}$, $AC = 6\,\text{cm}$ and $\hat{A} = 70°$. Find the remaining side BC.

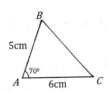

Solution. We are given $c = 5\,\text{cm}$, $b = 6\,\text{cm}$, $A = 70°$.

Fig. 4.3.1

$a = BC$ is found from
$$a^2 = 6^2 + 5^2 - 2 \times 6 \times 5 \times \cos 70°$$

$$= 61 - 60\cos 70° \simeq 40.48\,\text{cm}^2.$$

$$\therefore a \simeq 6.36\,\text{cm}.$$

Exercise. Having found BC in the above problem, find $\hat{B}$ by another application of the cosine rule.   ◇

Note: It is easier to do this with the help of the sine rule, in Chapter 6.

<u>Ans:</u> $\cos B \simeq 0.463$ then gives $B \simeq 62.4°$.

2) Refer to Fig. 4.3.2.

Find $\hat{A}$ in $\triangle ABC$ when $a = 7\,\text{cm}$, $b = 8\,\text{cm}$, $c = 9\,\text{cm}$.

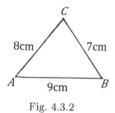

Fig. 4.3.2

Solution.

$$\cos A = \frac{b^2 + c^2 - a^2}{2bc}$$
$$= \frac{8^2 + 9^2 - 7^2}{2 \times 8 \times 9}$$
$$= \frac{96}{16 \times 9}$$
$$= \frac{2}{3}.$$

Inverting this relation (using the shift-inverse cos facility on the calculator) gives $A \simeq 48.2°$.

Exercise. In this problem, find $\cos C$, $\hat{C}$ and $\hat{B}$.

<u>Ans:</u> $\cos C = \frac{2}{7}$, $C \simeq 73.4°$, $B \simeq 58.4°$.   ◇

Exercise. In problem 2) of 3.4, find $O\hat{C}D$ and, directly from the cosine rule, find the distance OD, from marina to destination.   ◇

Chapter 5

Stewart's Theorem. Medians and the Centroid G

5.1 Introduction

Stewart's theorem and a corollary, Apollonius' theorem, follow exactly from the cosine rule in Chapter 4. They lead to a discussion about medians of a triangle, and the existence of the point G within the triangle. It is the centroid; its connection with other centres of a triangle is investigated later. See in particular Chapters 28, 29.

5.2 Stewart's Theorem [M. Stewart, 1746]

Theorem 5.1 (Stewart's Theorem). *If for a $\triangle ABC$, c_A is the length from vertex A to a point Z on BC dividing BC so $BZ = m$, $ZC = n$, then*

$$a\left(c_A^2 + mn\right) = mb^2 + nc^2.$$

Note: $AZ = c_A$ *is called a cevian.*

Proof. Refer to Fig. 5.2.1.

By the cosine rule applied to triangles AZB, AZC,

$c^2 = c_A^2 + m^2 - 2c_A m \cos A\hat{Z}B$, and

$b^2 = c_A^2 + n^2 - 2c_A n \cos A\hat{Z}C$.

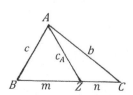

Fig. 5.2.1

Eliminating the trigonometric terms from the
two equations we find

$$mb^2 + nc^2 = (m+n)c_A^2 + mn^2 + nm^2$$
$$= (m+n)(c_A^2 + mn)$$
$$= a(c_A^2 + mn).$$

$\square$

Corollary 5.1. (Apollonius' Theorem [Apollonius of Perga, c. 262 BC–c. 190 BC])

If Z is the midpoint of BC, $AZ = c_A$ becomes the median, m_A.

Since $m = n = \frac{a}{2}$, Stewart's theorem yields

$$b^2 + c^2 = 2(m_A^2 + \tfrac{a^2}{4}) = 2m_A^2 + \tfrac{a^2}{2}.$$

Clearly also

$c^2 + a^2 = 2m_B^2 + \frac{b^2}{2}$ and $a^2 + b^2 = 2m_C^2 + \frac{c^2}{2}$ give the other two median
lengths m_B, m_C.

Exercise. Starting from the cosine rule, give a direct proof of Apollonius'
theorem.   ◇

5.3   Theorem

Theorem 5.2. *The medians AA', BB', CC' of $\triangle ABC$ are concurrent, at point G, called the centroid. It has the property that*

$$AG : GA' = BG : GB' = CG : GC' = 2 : 1, \ i.e.$$

$$AG = \frac{2m_A}{3},$$

$$BG = \frac{2m_B}{3},$$

$$CG = \frac{2m_C}{3}.$$

Proof. Refer to Fig. 5.3.1.

Let the medians AA', BB' meet at G. Draw $A'B'$.

$\because A'B' \parallel AB$ and $A'B' = \frac{AB}{2}$, $\triangle AGB$ and $\triangle A'GB'$ are similar.

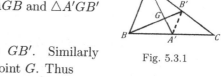

So $AG : GA' = 2 : 1 = BG : GB'$. Similarly AA', CC' will meet at the same point G. Thus

Fig. 5.3.1

$$AG = \frac{2AA'}{3},$$

$$BG = \frac{2BB'}{3},$$

$$CG = \frac{2CC'}{3}.$$

$\square$

5.4 Problems / exercises

1) Refer to Fig. 5.4.1.

Find the median lengths in the right-angled triangle with sides $3\,\text{cm}, 4\,\text{cm}, 5\,\text{cm}$.

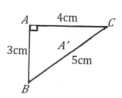

Solution. $AA' = 2.5\,\text{cm}$, obviously

Fig. 5.4.1

$$BB' = \sqrt{\frac{5^2+3^2-\frac{4^2}{2}}{2}} = \sqrt{\frac{25+9-8}{2}} = \sqrt{13}\,\text{cm}\ (\simeq 3.61\,\text{cm}),$$

$$CC' = \sqrt{\frac{5^2+4^2-\frac{3^2}{2}}{2}} = \sqrt{\frac{25+16-4.5}{2}} = \sqrt{18.25}\,\text{cm}\ (\simeq 4.27\,\text{cm}).$$

Exercise. Find the median lengths in a triangle having sides 7cm, 8cm, 9cm.

<u>Ans:</u> $7\,\text{cm}, 6.02\,\text{cm}, 7.76\,\text{cm}.$ $\diamond$

2) Show that $AG^2 + BG^2 + CG^2 = \frac{1}{3}(a^2 + b^2 + c^2).$

Solution. $AG = \frac{2}{3}m_A$, $m_A^2 = \frac{1}{2}(b^2 + c^2 - \frac{1}{2}a^2)$; with two pairs of similar equalities.

$$\therefore \quad AG^2 + BG^2 + CG^2$$

$$= \frac{4}{9}(m_A^2 + m_B^2 + m_C^2)$$

$$= \frac{2}{9}\{(b^2 + c^2 - \frac{1}{2}a^2) + (c^2 + a^2 - \frac{1}{2}b^2) + (a^2 + b^2 - \frac{1}{2}c^2)\}$$

$$= \frac{2}{9}\{2(a^2 + b^2 + c^2) - \frac{1}{2}(a^2 + b^2 + c^2)\}$$

$$= \frac{1}{3}(a^2 + b^2 + c^2).$$

Chapter 6

The Circumcentre, O. The Sine Rule

6.1 The circumcircle; centre O, radius R

Refer to Fig. 6.1.1.

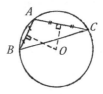

The circumcircle of $\triangle ABC$ has its centre at O. This
point is at the intersection of the perpendicular bisectors of the sides, viz. the chords BC, CA, AB of
the circle.

The radius of the circle is R, the circumradius, and
O is the circumcentre.

Fig. 6.1.1

6.2 The Sine Rule

Theorem 6.1 (The Sine Rule).
In $\triangle ABC$,

$$\frac{a}{\sin A} = \frac{b}{\sin B} = \frac{c}{\sin C} = 2R.$$

Or, equivalently,

$$a = 2R \sin A,$$
$$b = 2R \sin B,$$
$$c = 2R \sin C.$$

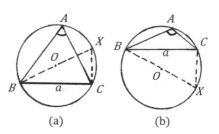

(a)   (b)

Fig. 6.2.1

Proof. Refer to Fig. 6.2.1.

In (a), A is acute; in (b), A is obtuse.

Draw the diameter BOX, and join X to C,

$$B\hat{X}C = \begin{cases} A & \text{in (a)}, \\ 180° - A & \text{in (b)}. \end{cases}$$

Since $\sin A = \sin(180° - A)$, from Theorem 3.1 part (iii),

$$a = BC = BX \sin B\hat{X}C = 2R\sin A.$$

In an identical manner, $b = 2R\sin B, c = 2R\sin C.$ □

The sine rule helps to solve a triangle.

6.3   Problems / exercises

1) Refer to Fig. 6.3.1.

In $\triangle ABC, B\hat{A}C = 110°, AB = 3\,\text{cm}, BC = 5\,\text{cm}$. Find the remaining angles of the triangle.

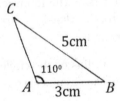

Fig. 6.3.1

Solution. By the sine rule,

$$\frac{5}{\sin 110°} = \frac{3}{\sin C}.$$

$\therefore \sin C = \frac{3}{5}\sin 110° \simeq 0.5638.$

Inverting, $C \simeq 34.3°$; hence $B = 180° - (A + C) \simeq 35.7°$.

Exercise. $\triangle ABC$ has $a = 10\,\text{cm}, B = 59°, C = 28°$. Find b, c.

Ans: $b = 8.58\,\text{cm}, c = 4.70\,\text{cm}$. (2 d.p.)   ◇

Chapter 7

Area. Hero's Formula

7.1 Notation

First, we let $s = \frac{1}{2}(a+b+c)$. This symbol is reserved for the semi-perimeter of $\triangle ABC$.

We put, also, Δ = area of $\triangle ABC$ and as before, R = circumradius of $\triangle ABC$.

This notation will be strictly adhered to in the subsequent work.

7.2 Theorem

Theorem 7.1. $\Delta = \dfrac{1}{2}bc\sin A$, $\left(or\ \dfrac{1}{2}ca\sin B, \dfrac{1}{2}ab\sin C\right)$

$$= \frac{abc}{4R}.$$

Proof. The height from B to AC is $h_B = c\sin A$, whether A is acute or obtuse. Therefore, $\Delta = \dfrac{1}{2}bh_B = \dfrac{1}{2}bc\sin A$.

(The other expressions in parentheses are found by cyclic interchanges of the letters.)

$$\Delta = \frac{1}{2}bc\sin A = \frac{1}{2}bc\frac{a}{2R}, \quad \text{from the sine rule of Theorem 6.1,}$$

$$= \frac{abc}{4R}.$$

$\square$

7.3   Hero's formula [Hero of Alexandria, c. 10–c. 70 AD]

Theorem 7.2 (Hero's Formula). *For any* $\triangle ABC$,
$$\Delta = \sqrt{s(s-a)(s-b)(s-c)}.$$

Proof.

$h_B^2 = (c\sin A)^2 = c^2(1 - \cos^2 A)$, from (i) of Theorem 3.1,

$\qquad = c^2\left\{1 - \left(\dfrac{b^2 + c^2 - a^2}{2bc}\right)^2\right\}$, from the cosine rule of Theorem 4.1,

$\qquad = c^2\left\{1 - \left(\dfrac{b^2 + c^2 - a^2}{2bc}\right)\right\}\left\{1 + \left(\dfrac{b^2 + c^2 - a^2}{2bc}\right)\right\}.$

(factorising the difference of squares.)

Hence,

$$h_B^2 = \frac{1}{4b^2}\left\{a^2 - (b-c)^2\right\}\left\{(b+c)^2 - a^2\right\}$$

$$= \frac{1}{4b^2}(a - b + c)(a + b - c)(b + c - a)(b + c + a),$$

(achieving a complete factorisation)

$\therefore\quad (2bh_B)^2 = (2s - 2b)(2s - 2c)(2s - 2a)(2s)$, on introducing the semi-perimeter s.

Dividing by 16: $\left(\frac{1}{2}bh_B\right)^2 = \Delta^2 = s(s-a)(s-b)(s-c)$. $\qquad\qquad\square$

7.4   Problems / exercises

1) Find the exact area of a triangle with sides $7\,\mathrm{cm}$, $8\,\mathrm{cm}$, $9\,\mathrm{cm}$. Find also the diameter of the circumscribing circle.

Solution. First, we find $s = \frac{1}{2}(7 + 8 + 9) = 12\,\mathrm{cm}$.

Then by Hero's formula,
$$\Delta = \sqrt{12(12 - 7)(12 - 8)(12 - 9)}$$
$$= \sqrt{12 \times 5 \times 4 \times 3}$$
$$= 12\sqrt{5}\,\mathrm{cm}^2.$$

Using the second expression for Δ in Theorem 7.1.

$$\text{diameter} = 2R = \frac{7 \times 8 \times 9}{2\Delta} = \frac{7 \times 36}{12\sqrt{5}} = \frac{21}{\sqrt{5}} \text{ cm.}$$

2) A parallelogram has two sides of 5 cm, 6 cm and an included angle of 57°. Find its area.

Solution. Using the first formula of Theorem 7.1.

$$\begin{aligned} \text{area of parallelogram} &= 2\left(\frac{1}{2} \times 5 \times 6 \times \sin 57°\right) \\ &= 30 \sin 57° \\ &\simeq 25.16 \text{ cm}^2. \end{aligned}$$

3) Using the details in the proof of Hero's formula, show that

$$16\Delta^2 = 2(a^2 b^2 + b^2 c^2 + c^2 a^2) - (a^4 + b^4 + c^4).$$

Solution.

$$\begin{aligned} 16\Delta^2 &= \left\{a^2 - (b-c)^2\right\}\left\{(b+c)^2 - a^2\right\} \\ &= -a^4 + a^2\left\{(b+c)^2 + (b-c)^2\right\} - (b^2 - c^2)^2, \text{ expanding out,} \\ &= -a^4 + 2a^2(b^2 + c^2) - (b^4 - 2b^2 c^2 + c^4) \\ &= 2(a^2 b^2 + b^2 c^2 + c^2 a^2) - (a^4 + b^4 + c^4), \text{ as required.} \end{aligned}$$

Exercise. Verify, arithmetically, the identity in this problem in the case of the right-angled triangle with sides 3 cm, 4 cm, 5 cm. ◇

Chapter 8

The Tangent Ratio

8.1 Definition

Of equal importance to the sine and cosine ratios is the tangent ratio. For an ordinary point plotted with coordinates (x, y), or $[r, \theta]$, we define tangent of θ, $\tan \theta = \frac{y}{x}$.

So, $\tan \theta = \dfrac{\frac{y}{r}}{\frac{x}{r}} = \dfrac{\sin \theta}{\cos \theta}$.

This ratio can assume any real value unless $x = 0$ (i.e. $\cos \theta = 0$) when the result is infinite.

E.g., $\tan 0 = \tan 180° = 0, \tan 45° = 1$, but $\tan 90°$ is infinite.

Exercise. On a calculator, check on the following values:

$$\tan 50° \simeq 1.918, \tan 91° \simeq -57.29, \tan -170° = -0.1783.$$

$\diamond$

8.2 Theorem

Theorem 8.1. (i) $\tan(-\theta) = -\tan \theta$,

(ii) $\tan(90° \pm \theta) = \mp \dfrac{1}{\tan \theta}$,

(iii) $\tan(180° \pm \theta) = \pm \tan \theta$,

unless in (ii) $\tan \theta = 0$ *(e.g.* $\theta = 0, 180°$*).*

Proof. Simple figures could be used to prove these, as was done for sines and cosines in the theorem in 3.3. But we can proceed in a different manner, and prove just a sample:

(i) $\tan(-\theta) = \frac{\sin(-\theta)}{\cos(-\theta)} = \frac{-\sin\theta}{\cos\theta}$, from properties already proved for sine and cosine.

$\therefore \tan(-\theta) = -\tan\theta.$

This shows that tangent is an odd function. See the remarks in 3.3.

(ii) (1st part): $\tan(90° + \theta) = \frac{\sin(90°+\theta)}{\cos(90°+\theta)} = \frac{+\cos\theta}{-\sin\theta} = -\frac{1}{\tan\theta}.$

(iii) (2nd part): $\tan(180° - \theta) = \frac{\sin(180°-\theta)}{\cos(180°-\theta)} = \frac{+\sin\theta}{-\cos\theta} = -\tan\theta.$ □

Exercise. Prove in an identical way the two remaining identities in the theorem.   ◇

8.3   Problems / exercises

1) Refer to Fig. 8.3.1.

A cliff OT is 100 m high. From the top T, two boats B_1, B_2 are observed. B_1 is North and B_2 is East of O. Their angles of depression from T are, respectively, 30°, 50°. Find the distance between B_1 and B_2.

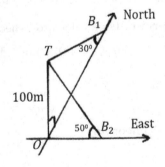

Fig. 8.3.1

Solution. $OT \perp OB_1$ and $OT \perp OB_2$.

$\frac{OT}{OB_1} = \tan 30°, \quad \therefore \quad OB_1 = \frac{100}{\tan 30°} \simeq 173$ m.

$\frac{OT}{OB_2} = \tan 50°, \quad \therefore \quad OB_2 = \frac{100}{\tan 50°} \simeq 84$ m.

Because $B_1\hat{O}B_2 = 90°$, $B_1B_2 \simeq \sqrt{173^2 + 84^2} \simeq 192$ m.

Exercise. Find the bearing of B_1 from B_2.

Ans: 334.1° (or N.25.9°W.)   ◇

2) Refer to Fig. 8.3.2.

A hillside slopes northwards. It is inclined
to the horizontal at an angle α. A straight
path up the hillside has a bearing of θ (i.e.,
N.θE.) and is inclined at an angle β to the
horizontal.

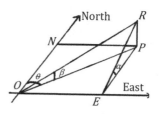

Show that $\tan\beta = \cos\theta\tan\alpha$.

Fig. 8.3.2

Solution. OR is the path up the hillside, RP is the height of R above O.

$N\hat{O}P = \theta$, the bearing of R from O;

$R\hat{E}P = \alpha$, the inclination of the hillside;

$P\hat{O}R = \beta$, the slope of the path.

$\cos\theta\tan\alpha = \frac{ON}{OP}\frac{PR}{PE} = \frac{PE}{OP}\frac{PR}{PE} = \frac{PR}{OP} = \tan\beta$, as required.

Exercise. Find β if $\theta = 60°$, $\alpha = 30°$ in the above.

Ans: 16.1°.   ◇

3) If $0 < \theta < 90°$ and $\tan\theta = k$, find $\cos\theta$ in terms of k.

Solution. $k = \tan\theta = \frac{\sin\theta}{\cos\theta} = \frac{\sqrt{1-\cos^2\theta}}{\cos\theta}$,

$\therefore k^2 = \frac{1}{\cos^2\theta} - 1$; so, $\cos\theta = \frac{1}{\sqrt{k^2+1}}$.

Exercise. If $90° < \theta < 180°$ and $\sin\theta = \frac{9}{41}$, find $\tan\theta$ (without finding θ first).

Ans: $-\frac{9}{40}$.   ◇

Chapter 9

Some Very Special Angles

Remark 9.1. A few special angles keep appearing in elementary problems, e.g. $30°$, $45°$, $60°$ and a number of others. The trigonometric ratios of such angles generally involve surd quantities and can be easily found from simple diagrams.

9.1 Ratios for $45°, 22.5°, 67.5°$

Refer to Fig. 9.1.1.

$\triangle ABC$ is an isosceles right-angled triangle, with $AB = BC = a$, $A\hat{B}C = 90°$. Then,

$AC = a\sqrt{2};$
$\sin 45° = \frac{AB}{AC} = \frac{1}{\sqrt{2}} = \cos 45°$, $\tan 45° = 1$.

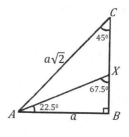

Fig. 9.1.1

Let AX bisect $B\hat{A}C$: $B\hat{A}X = 22.5°$ and $A\hat{X}B = 67.5°$.

Put $BX = x$.

Because $\frac{BX}{XC} = \frac{AB}{AC}$ ($\because AX$ is a bisector), $\frac{x}{a-x} = \frac{1}{\sqrt{2}}$,

$(\sqrt{2} + 1)x = a, x = \frac{a}{\sqrt{2}+1} = a(\sqrt{2} - 1).$

$\therefore \tan 22.5° = \frac{BX}{AB} = \frac{x}{a} = \sqrt{2} - 1$ and $\tan 67.5° = \frac{AB}{BX} = \frac{a}{x} = \sqrt{2} + 1.$

9.2   Ratios for $60°, 30°, 15°, 75°$

Refer to Fig. 9.2.1.

$\triangle ABC$ is equilateral, with $AB = BC = CA = 2a$. AM bisects the triangle.

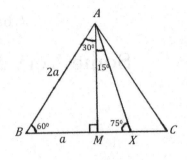

$BM = a, AM = a\sqrt{3}.$

$\therefore \sin 30° = \frac{BM}{AB} = \frac{1}{2} = \cos 60°.$

$\tan 30° = \frac{BM}{AM} = \frac{1}{\sqrt{3}};$

$\tan 60° = \frac{AM}{BM} = \sqrt{3}.$

Fig. 9.2.1

Let AX bisect $M\hat{A}C$: $M\hat{A}X = 15°$, $A\hat{X}M = 75°$.

Put $MX = x$ so $XC = a - x$

Because $\frac{MX}{XC} = \frac{AM}{AC}, \frac{x}{a-x} = \frac{\sqrt{3}}{2},$

$\sqrt{3}(a - x) = 2x, x = \frac{a\sqrt{3}}{2+\sqrt{3}} = a\sqrt{3}(2 - \sqrt{3}),$

$\tan 15° = \frac{MX}{AM} = \frac{x}{a\sqrt{3}} = 2 - \sqrt{3}$ and

$\tan 75° = \frac{AM}{MX} = \frac{1}{2-\sqrt{3}} = 2 + \sqrt{3}.$

9.3   Problems / exercises

1) From the results found in 9.2, find $\sin 67.5°$.

Solution. Put $x = \sin 67.5°$.

$$x^2 = \sin^2 67.5°$$
$$= (\tan 67.5° \cos 67.5°)^2$$
$$= (\sqrt{2} + 1)^2(1 - x^2)$$
$$= (3 + 2\sqrt{2})(1 - x^2).$$

$\therefore x^2(4 + 2\sqrt{2}) = 3 + 2\sqrt{2},$

$$x^2 = \frac{3 + 2\sqrt{2}}{2(2 + \sqrt{2})} = \frac{(3 + 2\sqrt{2})(2 - \sqrt{2})}{4} = \frac{2 + \sqrt{2}}{4}.$$

$$\therefore x = \frac{1}{2}\sqrt{2 + \sqrt{2}} = \sin 67.5^\circ.$$

Exercise. Show $\cos 67.5^\circ = \frac{1}{2}\sqrt{2 - \sqrt{2}}$. ◇

2) From the results in 9.2, find $\cos 15^\circ$.

Solution. Put $y = \cos 15^\circ$, $y^2 = \cos^2 15^\circ = \frac{\sin^2 15^\circ}{\tan^2 15^\circ} = \frac{1 - y^2}{(2 - \sqrt{3})^2}$,

$$(7 - 4\sqrt{3})y^2 = 1 - y^2, \quad y^2 = \frac{1}{4(2 - \sqrt{3})} = \frac{2 + \sqrt{3}}{4}.$$

$$\therefore y = \frac{1}{2}\sqrt{2 + \sqrt{3}} = \cos 15^\circ$$

Exercise. Show that $\cos 15^\circ$ can be re-written as $\frac{1}{4}(\sqrt{6} + \sqrt{2})$. ◇

9.4 Summary

Refer to Table 9.4.1. The results obtained above are contained therein. The remaining spaces have been filled in.

θ	$\sin \theta$	$\cos \theta$	$\tan \theta$
15°	$\frac{1}{4}(\sqrt{6} - \sqrt{2})$	$\frac{1}{4}(\sqrt{6} + \sqrt{2})$	$2 - \sqrt{3}$
22.5°	$\frac{1}{2}\sqrt{2 - \sqrt{2}}$	$\frac{1}{2}\sqrt{2 + \sqrt{2}}$	$\sqrt{2} - 1$
30°	$\frac{1}{2}$	$\frac{\sqrt{3}}{2}$	$\frac{1}{\sqrt{3}}$
45°	$\frac{1}{\sqrt{2}}$	$\frac{1}{\sqrt{2}}$	1
60°	$\frac{\sqrt{3}}{2}$	$\frac{1}{2}$	$\sqrt{3}$
67.5°	$\frac{1}{2}\sqrt{2 + \sqrt{2}}$	$\frac{1}{2}\sqrt{2 - \sqrt{2}}$	$\sqrt{2} + 1$
75°	$\frac{1}{4}(\sqrt{6} + \sqrt{2})$	$\frac{1}{4}(\sqrt{6} - \sqrt{2})$	$2 + \sqrt{3}$

Table 9.4.1

For an extended list of special values, see pp. 31, 74 in [1]. See also p. 35 in [7] for the special angles 18°, 36°, 54°, 72°.

Chapter 10

Cosecant, Secant, Cotangent. Proving Simple Identities

Remark 10.1. Three further ratios complete the set of trigonometric ratios needed in problem solving.

10.1 Definitions

$$\text{cosecant of } \theta, \quad \operatorname{cosec}\theta = \tfrac{1}{\sin\theta};$$
$$\text{secant of } \theta, \quad \sec\theta = \tfrac{1}{\cos\theta};$$
$$\text{cotangent of } \theta, \quad \cot\theta = \tfrac{1}{\tan\theta}.$$

These are all finite, except when the denominators on the right are zero.

Clearly, $|\operatorname{cosec}\theta| \geq 1$, $|\sec\theta| \geq 1$, while $\cot\theta$ can take all values. In some books, these ratios are abbreviated to csc, sec, ctg respectively. We do not use these.

Exercise. Confirm the following with your calculator:

$$\operatorname{cosec}46° \simeq 1.390, \quad \sec 105° \simeq -3.864, \quad \cot(-2°) \simeq -28.64.$$

◇

10.2 Theorem

Theorem 10.1. *For all* θ, (i) $1 + \cot^2\theta = \operatorname{cosec}^2\theta$,

(ii) $\tan^2\theta + 1 = \sec^2\theta$.

Proof. Begin with the basic identity $\sin^2\theta + \cos^2\theta = 1$.

(i) Divide this through by $\sin^2\theta$:

$$1 + \left(\frac{\cos\theta}{\sin\theta}\right)^2 = \left(\frac{1}{\sin\theta}\right)^2, \text{ so that } 1 + \cot^2\theta = \operatorname{cosec}^2\theta,$$

(ii) Instead, divide by $\cos^2\theta$:

$$\left(\frac{\sin\theta}{\cos\theta}\right)^2 + 1 = \left(\frac{1}{\cos\theta}\right)^2, \text{ so that } \tan^2\theta + 1 = \sec^2\theta. \qquad \square$$

10.3   Problems / exercises

1) If $(\sec\theta - \cos\theta)^2 + (\operatorname{cosec}\theta - \sin\theta)^2 = p$, and $\cos\theta\sin\theta = q$, prove that

$$pq^2 = 1 - 3q^2.$$

Solution.

$$p = (\sec^2\theta - 2 + \cos^2\theta) + (\operatorname{cosec}^2\theta - 2 + \sin^2\theta)$$
$$= \sec^2\theta + \operatorname{cosec}^2\theta - 3.$$

Hence,

$$pq^2 = (\sec^2\theta + \operatorname{cosec}^2\theta - 3)\cos^2\theta\sin^2\theta$$
$$= \sin^2\theta + \cos^2\theta - 3\cos^2\theta\sin^2\theta$$
$$= 1 - 3q^2.$$

2) If $\operatorname{cosec}\theta + \sec\theta = \sqrt{2}$, show that $\cos^3\theta + \sin^3\theta = -\frac{1}{\sqrt{2}}$.

Solution. We are, effectively, given that $\dfrac{1}{\sin\theta} + \dfrac{1}{\cos\theta} = \sqrt{2}$, or $\cos\theta + \sin\theta = \sqrt{2}\cos\theta\sin\theta$.

Then

$$\cos^3\theta + \sin^3\theta = (\cos\theta + \sin\theta)(\cos^2\theta - \cos\theta\sin\theta + \sin^2\theta),$$

on factorising the sum of cubes.

$\therefore \quad \cos^3\theta + \sin^3\theta = (\cos\theta + \sin\theta)(1 - \cos\theta\sin\theta)$

$$= \cos\theta + \sin\theta - \frac{1}{\sqrt{2}}(\cos\theta + \sin\theta)^2$$

$$= \cos\theta + \sin\theta - \frac{1}{\sqrt{2}}(1 + 2\cos\theta\sin\theta)$$

$$= -\frac{1}{\sqrt{2}} + (\cos\theta + \sin\theta - \sqrt{2}\cos\theta\sin\theta)$$

$$= -\frac{1}{\sqrt{2}}.$$

Exercise. Verify that

(a) $\cot\theta + \tan\theta = \operatorname{cosec}\theta\sec\theta$;

(b) $(\operatorname{cosec}\theta + \cot\theta)^2 = \dfrac{1 + \cos\theta}{1 - \cos\theta}$;

(c) $\tan^2\theta + \tan^4\theta = \sec^4\theta - \sec^2\theta$.

$\diamond$

Chapter 11

Further Problems — Heights & Distances

11.1 Problems

1) Refer to Fig. 11.1.1.

A regular pyramid of height h has a square base of side a. Show that the angle between adjacent triangular sloping faces is 2θ, where

$$\sin\theta = \sqrt{\frac{a^2 + 2h^2}{a^2 + 4h^2}}.$$

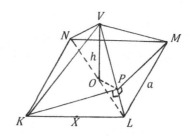

Fig. 11.1.1

Solution. V is the vertex, $KLMN$ is the square base, side a; its centre is O. $h = OV \perp$ base.

Let X be the midpoint of KL.

$OK = \frac{a}{\sqrt{2}} = OL, OX = \frac{a}{2}$, so that

$VX = \sqrt{\frac{a^2}{4} + h^2}, VL = \sqrt{\frac{a^2}{2} + h^2}.$

Choose P on VL so $KP \perp VL$ and $MP \perp V$.

OP, being on the plane VOL, we know $OK \perp OP$.

The angle required is that between (e.g.) triangular faces VKL, VLM.

It is $K\hat{P}M = 2O\hat{P}K = 2\theta$, with the notation in the question.

We find KP from $\frac{1}{2}VX \cdot KL = \text{area} \triangle VKL = \frac{1}{2}KP \cdot VL.$

Thus

$$KP = \frac{VX \cdot a}{VL} = \sqrt{\frac{\frac{a^2}{4} + h^2}{\frac{a^2}{2} + h^2}}.$$

$$\sin O\hat{P}K = \sin\theta = \frac{\frac{a}{\sqrt{2}}}{KP} = \frac{1}{\sqrt{2}}\sqrt{\frac{\frac{a^2}{2} + h^2}{\frac{a^2}{4} + h^2}} = \sqrt{\frac{a^2 + 2h^2}{a^2 + 4h^2}},$$

whence θ and 2θ.

Exercise.

(i) In the above, show that the angle between opposite triangular faces is 2φ, where $\tan\varphi = \frac{a}{2h}$.

(ii) The Great Pyramid of Cheops at Giza, near Cairo, is 147 m high, and is on a square base of side 230 m. What is

    a) the angle between adjacent triangular faces?
    b) the angle between opposite triangular faces?

<u>Ans:</u> a) 112°, b) 76°.

                                    ◇

2) Refer to Fig. 11.1.2.

A tower of height h stands on a hill which slopes uniformly at an angle α to the horizontal. A boy stands at the foot of the hill on the line of greatest slope of the tower notices that the tower is due North of him.

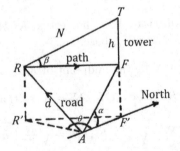

Fig. 11.1.2

On ascending a distance d along the straight road up the hill in a direction θ west of north, the boy comes to a horizontal path on the same level as, and leading to, the foot of the tower. At this point he finds the angle of elevation of the tower is β.

Prove that

$$h = \frac{d\tan\beta}{\sqrt{1 + \cot^2\theta\sec^2\alpha}}.$$

Solution. FT is the tower, height h.

A is the boy's starting position.

AR is the road uphill, RF is the horizontal path to the tower.

F', R' are the projections of F, R at the level of A.

$F\hat{A}F' = \alpha$,

$R'\hat{A}F' = \theta$, the westerly bearing of the road.

$T\hat{R}F = \beta$.

We have $FR = FT\cot\beta = h\cot\beta, AR = d$.

$\therefore$ (i) $AF^2 = d^2 - h^2\cot^2\beta$,

$$\tan\theta = \frac{F'R'}{AF'} = \frac{FR}{AF\cos\alpha} = \frac{h\cot\beta}{AF\cos\alpha}.$$

$\therefore$ (ii) $AF = h\cot\beta\sec\alpha\cot\theta$.

Now we eliminate AF between (i) and (ii) to give

$$d^2 - h^2\cot^2\beta = h^2\cot^2\beta\sec^2\alpha\cot^2\theta,$$

$$d^2 = h^2\cot^2\beta(1 + \cot^2\theta\sec^2\alpha).$$

$\therefore \quad d\tan\beta = h\sqrt{1 + \cot^2\theta\sec^2\alpha}$ whence h.

Exercise. If in the above problem $\theta = 135°$, $\alpha = 45°$, $\beta = 30°$, show that $h = \frac{d}{3}$.

$\diamond$

Chapter 12

The Factorisation Formulae & Napier's Tangent Rule

12.1 Factorisation formulae

Theorem 12.1 (The Factorisation Formulae). *For any angles P, Q,*

$$\text{(F1):}\quad \sin P + \sin Q = 2\sin \frac{P+Q}{2}\cos \frac{P-Q}{2};$$

$$\text{(F2):}\quad \sin P - \sin Q = 2\cos \frac{P+Q}{2}\sin \frac{P-Q}{2};$$

$$\text{(F3):}\quad \cos P + \cos Q = 2\cos \frac{P+Q}{2}\cos \frac{P-Q}{2};$$

$$\text{(F4):}\quad \cos P - \cos Q = -2\sin \frac{P+Q}{2}\sin \frac{P-Q}{2}.$$

Proof. Refer to Fig. 12.1.1.

A rhombus $OSTR$ is drawn with sides $= 1$. $R\hat{O}x = P$, $S\hat{O}x = Q$. Diagonals OT, RS bisect each other at right angles, at the point M. We put P, Q in the 1st quadrant but this is not essential to the proof.

Let $\frac{P+Q}{2} = A, \frac{P-Q}{2} = B$. Then OT makes $\hat{A}$ with Ox and RS makes $\hat{A}$ with Oy.

Also, $R\hat{O}M = M\hat{O}S = B$.

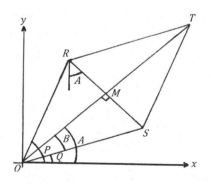

Fig. 12.1.1

51

Since $OS = 1, OM = \cos B, MS = \sin B$.

Thus

$$\sin P + \sin Q = y\text{-coord. of } R + y\text{-coord. of } S$$
$$= 2 \cdot y\text{-coord. of } M$$
$$= 2 \cdot OM \sin A$$
$$= 2 \sin A \cos B. \text{ This is (F1)}$$
$$\cos P - \cos Q = x\text{-coord. of } R - x\text{-coord. of } S$$
$$= -2 \cdot MS \sin A$$
$$= -2 \sin A \sin B. \text{ This is (F4)}$$

(F2), (F3) are similarly proved.   □

Exercise. Prove (F2) and (F3).   ◇

12.2   Napier's Tangent Rule [J. Napier, 1550–1617]

Theorem 12.2 (Napier's Tangent Rule). *In a $\triangle ABC$,*

$$\tan \frac{A-B}{2} = \frac{a-b}{a+b} \Big/ \tan \frac{C}{2}.$$

Note: This formula was published by Napier in 1614 in his book on logarithms.

Proof. From the sine rule of Theorem 6.1,

$$\frac{a-b}{a+b} = \frac{2R(\sin A - \sin B)}{2R(\sin A + \sin B)}$$
$$= \frac{2 \cos \frac{A+B}{2} \sin \frac{A-B}{2}}{2 \sin \frac{A+B}{2} \cos \frac{A-B}{2}}, \text{ by (F1) and (F2).}$$

$$\therefore \quad \frac{a-b}{a+b} = \frac{\tan \frac{A-B}{2}}{\tan \frac{A+B}{2}}.$$

But $\tan \dfrac{A+B}{2} = \tan\left(90° - \dfrac{C}{2}\right) = \cot \dfrac{C}{2} = 1 \Big/ \tan \dfrac{C}{2}$. The result follows.

□

12.3 Problems / exercises

1) Evaluate $\dfrac{\sin 15\theta - \sin 7\theta}{\sin 6\theta + \sin 2\theta}$ when $\theta = 20°$.

Solution.

$$\text{The ratio} \; = \frac{2\cos 11\theta \sin 4\theta}{2\sin 4\theta \cos 2\theta}, \text{ from (F1) and (F2)},$$
$$= \frac{\cos 11\theta}{\cos 2\theta}.$$

When $\theta = 20°$, this is $\dfrac{\cos 220°}{\cos 40°} = \dfrac{\cos(180° + 40°)}{\cos 40°} = \dfrac{-\cos 40°}{\cos 40°} = -1.$

2) Show that $\cos 50° + \cos 10° + \cos 40° + \cos 20° = 2\sqrt{3}\cos 15° \cos 5°$.

Solution.

$$\text{LHS} \; = 2\cos 30° \cos 20° + 2\cos 30° \cos 10°, \text{ using (F3) twice},$$
$$= \sqrt{3}(\cos 20° + \cos 10°), \text{ since } \cos 30° = \frac{\sqrt{3}}{2},$$
$$= 2\sqrt{3}\cos 15° \cos 5°, \text{ on using (F3) again},$$
$$= \text{RHS}.$$

3) In $\triangle ABC, a = 9\,\text{cm}, b = 7\,\text{cm}, C = 80°$; find A, B by Napier's rule.

Solution. By Napier's rule of Theorem 12.2,

$$\tan\frac{A-B}{2} = \frac{9-7}{9+7}\bigg/ \tan\frac{80°}{2} = \frac{\frac{1}{8}}{\tan 40°} \simeq 0.1490.$$

From this $\frac{A-B}{2} \simeq 8.475°$, $A - B \simeq 16.95°$.

But $A + B = 180° - C = 100°$. Solving, $A \simeq 58.5°$, $B \simeq 41.5°$.

Exercise. With the data here, find c by the cosine rule, then A and B by the sine rule, and so verify the results in the solution. ◇

Chapter 13

Addition Formulae for Sines & Cosines

Remark On a par with the four factorisation formulae in 12.1., we can produce four addition (or "compound angle") formulae, as below.

13.1 Addition formulae

Theorem 13.1. *For any angles A, B,*

$$(A1):\quad \sin(A + B) = \sin A \cos B + \cos A \sin B;$$
$$(A2):\quad \sin(A - B) = \sin A \cos B - \cos A \sin B;$$
$$(A3):\quad \cos(A + B) = \cos A \cos B - \sin A \sin B;$$
$$(A4):\quad \cos(A - B) = \cos A \cos B + \sin A \sin B.$$

Proof. Consider (F1)–(F4) in Theorem 12.1.

Let $\frac{P+Q}{2} = A$, $\frac{P-Q}{2} = B$.

Thus $A+B = P$ and $A-B = Q$. Adding (F1) to (F2) will give $\sin(A+B) = \sin A \cos B + \cos A \sin B$. This is (A1).

Subtracting (F2) from (F1) gives (A2) just as easily. Otherwise, changing B to $-B$ in (A1) gives (A2), since $\cos(-B) = \cos B$ and $\sin(B) = -\sin B$.

(A3) and (A4) come from (F3), (F4) in exactly the same way. $\qquad\square$

Exercise. Prove (A3), (A4). $\qquad\qquad\diamond$

Corollary 13.1. *In (A1), (A3) put $B = A = \theta$ or $\frac{\theta}{2}$ to give these important special cases:*

$$\begin{cases} \sin 2\theta = 2\sin\theta\cos\theta; \\[1em] \cos 2\theta = \cos^2\theta - \sin^2\theta \\ \qquad = 2\cos^2\theta - 1 \\ \qquad = 1 - 2\sin^2\theta; \end{cases} \quad or \quad \begin{cases} \sin\theta = 2\sin\frac{\theta}{2}\cos\frac{\theta}{2}; \\[1em] \cos\theta = \cos^2\frac{\theta}{2} - \sin^2\frac{\theta}{2} \\ \qquad = 2\cos^2\frac{\theta}{2} - 1 \\ \qquad = 1 - 2\sin^2\frac{\theta}{2}. \end{cases}$$

13.2   Problems / exercises

1) Show that $\sin 3\theta = 3\sin\theta - 4\sin^3\theta$.

Solution.

$$\sin 3\theta = \sin(2\theta + \theta) = \sin 2\theta\cos\theta + \cos 2\theta\sin\theta, \text{ by (A4)},$$
$$= 2\sin\theta\cos^2\theta + (1 - 2\sin^2\theta)\sin\theta, \text{ from the corollary,}$$
$$= 2\sin\theta(1 - \sin^2\theta) + (1 - 2\sin^2\theta)\sin\theta$$
$$= 3\sin\theta - 4\sin^3\theta, \text{ as required.}$$

Exercise. Similarly show that $\cos 3\theta = 4\cos^3\theta - 3\cos\theta$.   ◇

2) Find $\cos 15°(= \sin 75°)$ from an addition formula.

Solution.

$$\cos 15° = \cos(45° - 30°) = \cos 45°\cos 30° + \sin 45°\sin 30°, \text{ from (A4)}$$
$$= \frac{1}{\sqrt{2}}\frac{\sqrt{3}}{2} + \frac{1}{\sqrt{2}}\frac{1}{2}, \text{ using the special ratios in 9.1 and 9.2,}$$
$$= \frac{1}{2\sqrt{2}}(\sqrt{3} + 1)$$
$$= \frac{1}{4}(\sqrt{6} + \sqrt{2}).$$

This confirms a result in Table 9.4.1.

3) Show that $\cot\theta - \cot 2\theta = \operatorname{cosec} 2\theta$.

Solution.

$$\begin{aligned}
\cot\theta - \cot 2\theta &= \frac{\cos\theta}{\sin\theta} - \frac{\cos 2\theta}{\sin 2\theta} \\
&= \frac{\cos\theta}{\sin\theta} - \frac{2\cos^2\theta - 1}{2\sin\theta\cos\theta}, \text{ using Corollary 13.1,} \\
&= \frac{2\cos^2\theta - (2\cos^2\theta - 1)}{2\sin\theta\cos\theta} \\
&= \frac{1}{\sin 2\theta} \\
&= \operatorname{cosec} 2\theta.
\end{aligned}$$

Exercise. By repeated use of this equation, show that

$$\operatorname{cosec} 2\theta + \operatorname{cosec} 4\theta + \operatorname{cosec} 8\theta + \operatorname{cosec} 16\theta = \cot\theta - \cot 16\theta.$$

◇

4) From the relation $\dfrac{\cos(\alpha - \theta)}{\sin\alpha} = \dfrac{\cos(\beta + \theta)}{\sin\beta}$, prove that

$$2\tan\theta = \cot\beta - \cot\alpha.$$

Solution. Addition formulae (A3) and (A4) give

$$\frac{\cos\alpha\cos\theta + \sin\alpha\sin\theta}{\sin\alpha} = \frac{\cos\beta\cos\theta - \sin\beta\sin\theta}{\sin\beta}.$$

$$\cot\alpha\cos\theta + \sin\theta = \cot\beta\cos\theta - \sin\theta,$$

$$2\sin\theta = \cot\beta\cos\theta - \cot\alpha\cos\theta,$$

dividing by $\cos\theta$ on both sides gives

$$2\tan\theta = \cot\beta - \cot\alpha.$$

5) In a $\triangle ABC$ show that

(i) $s = 4R\cos\frac{A}{2}\cos\frac{B}{2}\cos\frac{C}{2}$;

(ii) $1 + \cos 2A + \cos 2B + \cos 2C = -4\cos A\cos B\cos C.$

Solution.

(i) $s = \dfrac{1}{2}(a+b+c) = R(\sin A + \sin B + \sin C)$,

by the sine rule (Theorem 6.1),

$$= R\left(2\sin\frac{A+B}{2}\cos\frac{A-B}{2} + 2\sin\frac{C}{2}\cos\frac{C}{2}\right),$$

using (F1) of Theorem 12.1 and Corollary 13.1,

$$= 2R\cos\frac{C}{2}\left(\cos\frac{A-B}{2} + \cos\frac{A+B}{2}\right), \text{ since } \frac{A+B}{2} = 90° - \frac{C}{2},$$

$$= 4R\cos\frac{C}{2}\cos\frac{A}{2}\cos\frac{B}{2}, \text{ by (F3) of Theorem 12.1.}$$

(ii) $1 + \cos 2A + \cos 2B + \cos 2C$

$$= 2\cos^2 A + 2\cos(B+C)\cos(B-C),$$

by Corollary 13.1 and (F3) of Theorem 12.1,

$$= 2\cos A(\cos A - \cos(B-C)), \text{ since } \cos(B+C) = -\cos A,$$

$$= -2\cos A(\cos(B+C) + \cos(B-C))$$

$$= -4\cos A\cos B\cos C.$$

Exercise. In $\triangle ABC$, prove that

$$\sin 2A + \sin 2B + \sin 2C = 4R\sin A\sin B\sin C.$$

◇

6) Refer to Fig. 13.2.1.

A ship is observed from a look-out tower at height h above sea-level. When the ship is closest to the tower, the ship's angle of depression from the tower is found to be α, and after time t it is β.

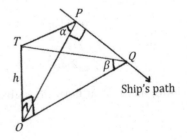

Fig. 13.2.1

Assuming uniform linear motion from P to Q, show that the ship's speed is

$$\frac{h}{t}\frac{\sqrt{\sin(\alpha+\beta)\sin(\alpha-\beta)}}{\sin\alpha\sin\beta}.$$

Solution. The ship's path is shown. OT is the tower at height h. P is the position of closest approach to the tower and Q is the position after time t.

$O\hat{P}Q = 90°, T\hat{P}O = \alpha, T\hat{Q}O = \beta$.

$OP = h\cot\alpha, OQ = h\cot\beta$,

so $PQ = \sqrt{OQ^2 - OP^2} = h\sqrt{\cot^2\beta - \cot^2\alpha}$.

Now,

$$\cot\beta - \cot\alpha = \frac{\cos\beta}{\sin\beta} - \frac{\cos\alpha}{\sin\alpha} = \frac{\sin\alpha\cos\beta - \cos\alpha\sin\beta}{\sin\alpha\sin\beta}$$

$$= \frac{\sin(\alpha - \beta)}{\sin\alpha\sin\beta}, \text{ on using (A2)},$$

and similarly, $\cot\beta + \cot\alpha = \dfrac{\sin(\alpha + \beta)}{\sin\alpha\sin\beta}$.

Thus, $PQ = h\sqrt{(\cot\beta - \cot\alpha)(\cot\beta + \cot\alpha)}$

$$= \frac{h\sqrt{\sin(\alpha + \beta)\sin(\alpha - \beta)}}{\sin\alpha\sin\beta}.$$

Dividing this by t gives the ship's speed.

Chapter 14

Addition Formulae for Tangents

14.1 Theorem

Theorem 14.1. $\tan(A \pm B) = \dfrac{\tan A \pm \tan B}{1 \mp \tan A \tan B}.$

Proof. From the addition formulae in Theorem 13.1, we have

$$\tan(A \pm B) = \frac{\sin(A \pm B)}{\cos(A \pm B)} = \frac{\sin A \cos B \pm \cos A \sin B}{\cos A \cos B \mp \sin A \sin B}.$$

On dividing numerator and denominator on the RHS by $\cos A \cos B$ we obtain

$$\tan(A \pm B) = \frac{\tan A \pm \tan B}{1 \mp \tan A \tan B}, \text{ as required.} \qquad \square$$

Corollary 14.1. *Putting $B = A$ in Theorem 14.1:* $\tan 2A = \dfrac{2 \tan A}{1 - \tan^2 A}$;

or letting $A = B = \frac{\theta}{2}$ in Theorem 14.1: $\tan \theta = \dfrac{2 \tan \frac{\theta}{2}}{1 - \tan^2 \frac{\theta}{2}}.$

Write this as $\tan \theta = \dfrac{2t}{1 - t^2}.$ *Then* $\sin \theta = \dfrac{2t}{\sqrt{(2t)^2 + (1 - t^2)^2}} =$

$\dfrac{2t}{\sqrt{1 + 2t^2 + t^4}} = \dfrac{2t}{1 + t^2}$ *and* $\cos \theta = \dfrac{1 - t^2}{\sqrt{(2t)^2 + (1 - t^2)^2}} = \dfrac{1 - t^2}{1 + t^2}.$

Exercise. Show that $\cot(A \pm B) = \dfrac{\cot A \cot B \mp 1}{\cot B \pm \cot A}.$ $\diamond$

61

14.2   Problems / exercises

1) Two lines on a plane have gradients $\frac{1}{2}$ and $\frac{1}{3}$. Find the angle between these lines.

Solution. Let the lines make angles θ, φ with Ox: $\tan\theta = \frac{1}{2}$, $\tan\varphi = \frac{1}{3}$

$\theta > \varphi$; and $\theta - \varphi$ is the angle required.

$$\tan(\theta - \varphi) = \frac{\frac{1}{2} - \frac{1}{3}}{1 + \frac{1}{2}\frac{1}{3}} = \frac{1}{7}; \theta - \varphi \simeq 8.1°.$$

2) If θ is acute and $\tan\theta = \frac{4}{3}$, show that $\tan\frac{\theta}{2} = \frac{1}{2}$.

Solution. By Corollary 14.1, $\dfrac{4}{3} = \tan\theta = \dfrac{2\tan\frac{\theta}{2}}{1 - \tan^2\frac{\theta}{2}}$,

$$\therefore 2\tan^2\frac{\theta}{2} + 3\tan\frac{\theta}{2} - 2 = 0.$$

$$\left(2\tan\frac{\theta}{2} - 1\right)\left(\tan\frac{\theta}{2} + 2\right) = 0.$$

Since θ is acute the only solution is $\tan\frac{\theta}{2} = \frac{1}{2}$.

Exercise. In problem 2 above, show that $\tan\frac{\theta}{4} = \sqrt{5} - 2$. $\diamond$

3) If $\cos\theta = \dfrac{5\cos\varphi - 3}{5 - 3\cos\varphi}$, prove that $\tan\dfrac{\theta}{2} = \pm 2\tan\dfrac{\varphi}{2}$.

Solution. Put $t = \tan\frac{\theta}{2}$, $T = \tan\frac{\varphi}{2}$.

We obtain $\dfrac{1 - t^2}{1 + t^2} = \dfrac{\frac{5(1-T^2)}{1+T^2} - 3}{5 - 3\frac{1-T^2}{1+T^2}} = \dfrac{1 - 4T^2}{1 + 4T^2}$, on simplifying.

This reduces easily to $t^2 = 4T^2$, therefore $t = \pm 2T$.

4) If A, B, C are the angles of a triangle, show that

$$\tan A + \tan B + \tan C = \tan A \tan B \tan C.$$

Solution. From the formula for $\tan(A+B)$ in Theorem 14.1,

$$\tan A + \tan B + \tan C = \tan(A+B)(1 - \tan A \tan B) + \tan C$$
$$= -\tan C(1 - \tan A \tan B) + \tan C,$$
$$\text{since } A + B = 180° - C,$$
$$= \tan A \tan B \tan C.$$

Exercise. (i) By applying Theorem 14.1 twice, show that

$$\tan(A+B+C) = \frac{\tan A + \tan B + \tan C - \tan A \tan B \tan C}{1 - (\tan B \tan C + \tan C \tan A + \tan A \tan B)}.$$

(ii) If $A + B + C = 90°$, find a relation between $\cot A$, $\cot B$, $\cot C$.
Ans: $\cot A \cot B \cot C = \cot A + \cot B + \cot C$. $\diamond$

Chapter 15

Further Half-Angle Formulae

15.1 Theorem

Theorem 15.1. *In* $\triangle ABC$, $\quad$ (i) $\sin \dfrac{A}{2} = \sqrt{\dfrac{(s-b)(s-c)}{bc}}$;

$$\text{(ii)} \quad \cos \frac{A}{2} = \sqrt{\frac{s(s-a)}{bc}};$$

$$\text{(iii)} \quad \tan \frac{A}{2} = \sqrt{\frac{(s-b)(s-c)}{s(s-a)}}.$$

Proof. (i) $2\sin^2\dfrac{A}{2} = 1 - \cos A$, from Corollary 13.1,

$$= 1 - \left(\frac{b^2 + c^2 - a^2}{2bc} \right), \text{ by the cosine rule,}$$

$$= \frac{1}{2bc}\{a^2 - (b-c)^2\}$$

$$= \frac{1}{2bc}(a + b - c)(a - b + c)$$

$$= \frac{(2s - 2c)(2s - 2b)}{2bc}$$

$$\therefore \sin^2 \frac{A}{2} = \frac{(s-b)(s-c)}{bc}.$$

(ii) is done similarly; or from (i) through $\cos^2 \dfrac{A}{2} = 1 - \sin^2\dfrac{A}{2}$.

(iii) follows easily by division. $\qquad\qquad\qquad\qquad\qquad\qquad$ □

15.2   Theorem

Theorem 15.2. *If AX bisects $\hat{A}$ of $\triangle ABC$, $AX = \dfrac{2\sqrt{bcs(s-a)}}{(b+c)}$.*

Proof. Refer to Fig. 15.2.1.

From $\frac{BX}{XC} = \frac{c}{b}$, we find $BX = \frac{ca}{(b+c)}$.

By applying the sine rule on $\triangle ABX$, we have

$$AX = \sin B \frac{BX}{\sin \frac{A}{2}} = \frac{ac \sin B}{b+c} \frac{1}{\sin \frac{A}{2}}$$

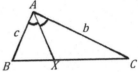

$$= \frac{2\Delta}{b+c} \Big/ \sqrt{\frac{(s-b)(s-c)}{bc}}$$

$$= \frac{2\sqrt{bcs(s-a)}}{(b+c)}, \text{ by Hero's formula.}$$

Fig. 15.2.1

$\square$

15.3   Problems / exercises

1) $\triangle ABC$ has $a = 7\,\text{cm}$, $b = 8\,\text{cm}$, $c = 9\,\text{cm}$. Find A.

Solution. $s = \frac{1}{2}(7 + 8 + 9) = 12\,\text{cm}$,

$\cos \frac{A}{2} = \sqrt{\frac{12(12-7)}{8 \times 9}} = \sqrt{\frac{5}{6}}$, from part (ii) of Theorem 15.1.

This gives $\frac{A}{2} \simeq 24.1°$, $A \simeq 48.2°$.

Exercise. Find B and C in the same way for this problem. Check using the cosine rule.

<u>Ans:</u> $B \simeq 58.4°$, $C \simeq 73.4°$. $\diamond$

2) Show that the length of the bisector of the right angle in the triangle with sides $3\,\text{cm}$, $4\,\text{cm}$, $5\,\text{cm}$ is exactly $\frac{12\sqrt{2}}{7}\,\text{cm}$.

Solution. $s = 6\,\text{cm}$; the required length $= \dfrac{2\sqrt{3\times4\times6(6-5)}}{(3+4)} = \dfrac{2\sqrt{72}}{7} = \dfrac{12\sqrt{2}}{7}\,\text{cm}$.

Exercise. Prove the Hero's formula of Theorem 7.2 from

$$\Delta = \frac{1}{2}bc\left(2\sin\frac{A}{2}\cos\frac{A}{2}\right).$$

◇

Chapter 16

Solving the Equation
$a \sin \theta + b \cos \theta = c$

Remark

The equation $a \sin \theta + b \cos \theta = c$, where a, b, c are constants can be solved for θ on a domain such as $-180° \leqslant \theta \leqslant 180°$, only provided that $|c| \leqslant \sqrt{a^2 + b^2}$. It will be shown below by essentially two methods. One depends on Theorem 13.1; the other on Corollary 14.1.

16.1 The auxiliary angle method

We need to solve $a \sin \theta + b \cos \theta = c$, if possible. There can be no solution if, for instance, $|c| > |a| + |b|$, for it is necessary to have $|c| = |a \sin \theta + b \cos \theta| \leqslant |a| + |b|$. As a first step, we write

$$a \sin \theta + b \cos \theta = R\left(\sin \theta \frac{a}{R} + \cos \theta \frac{b}{R} \right),$$

where we choose $R = \sqrt{a^2 + b^2}$, so as to make

$$\left(\frac{a}{R} \right)^2 + \left(\frac{b}{R} \right)^2 = 1.$$

In consequence we can put, for instance, $\frac{a}{R} = \sin \alpha$, $\frac{b}{R} = \cos \alpha$. This will define α as an angle in one of the 4 quadrants depending on the signs of a, b. α is known as an auxiliary angle. Then

$$a \sin \theta + b \cos \theta = R(\sin \theta \sin \alpha + \cos \theta \cos \alpha) \equiv R \cos(\theta - \alpha)$$

by an addition formula. The given equation is finally reduced to

$$\cos(\theta - \alpha) = \frac{c}{\sqrt{a^2 + b^2}}.$$

69

The last step determines from this, a number of possibilities for $(\theta - \alpha)$ and hence for θ, provided only that $|c| \leq R = \sqrt{a^2 + b^2}$.

Note: Any of the forms $R\cos(\theta - \alpha)$, $R\cos(\theta + \alpha)$, $R\sin(\theta - \alpha)$, $R\sin(\theta + \alpha)$ can be chosen to effect a solution.

16.2   Theorem

Theorem 16.1. *The equation* $a\sin\theta + b\cos\theta = c$ *has solutions for* θ, *given by*

$$\tan\frac{\theta}{2} = \frac{a \pm \sqrt{a^2 + b^2 - c^2}}{b + c}$$

provided $c^2 \leq a^2 + b^2$.

Proof. Put $\sin\theta = \frac{2t}{1+t^2}$, $\cos\theta = \frac{1-t^2}{1+t^2}$, where $t \equiv \tan\frac{\theta}{2}$. (Corollary 14.1)

$$a\left(\frac{2t}{1+t^2}\right) + b\left(\frac{1-t^2}{1+t^2}\right) = c,$$

$$(b+c)t^2 - 2at + (c-b) = 0.$$

In the case $c = -b$, then $t = \frac{c-b}{2a}$ (assuming $a \neq 0$) and θ can be found. More usually $b + c \neq 0$ and the equation is quadratic. Its roots will be real only when the discriminant is not negative, i.e., when $a^2 \geqslant (c-b)(c+b) = c^2 - b^2$, i.e., $c^2 \leqslant a^2 + b^2$. Then

$$t = \tan\frac{\theta}{2} = \frac{a \pm \sqrt{a^2 + b^2 - c^2}}{b + c},$$

whence θ.   □

16.3   Problems / exercises

1) Apply the two methods described to solve the equation

$$6\sin\theta + 8\cos\theta = 5,$$

for θ in the range $-180° \leq \theta \leq 180°$.

Solution 1. Divide through by $\sqrt{6^2+8^2} = 10$,

$$0.6\sin\theta + 0.8\cos\theta = 0.5,$$

$\therefore \cos(\theta - \alpha) = 0.5$, where $\sin\alpha = 0.6, \cos\alpha = 0.8$ and α is in the 1st quadrant: $\alpha \simeq 36.87°$. Since $-180° \le \theta \le 180°$,

$$-216.87° = -180° - 36.87° \le \theta - \alpha \le 180° - 36.87° = 143.13,$$

$\therefore \theta - \alpha = 60°, -60°$.

Hence $\theta \simeq 96.9°, -23.1°$ are the required solutions.

Solution 2. The substitution $t = \tan\frac{\theta}{2}$ converts the equation to

$$6\left(\frac{2t}{1+t^2}\right) + 8\left(\frac{1-t^2}{1+t^2}\right) = 5,$$

$$13t^2 - 12t - 3 = 0,$$

$$\tan\frac{\theta}{2} \equiv t = \frac{6 \pm \sqrt{6^2 + 3 \times 13}}{13} = \frac{6 \pm \sqrt{75}}{13} \simeq 1.1277 \text{ or } -0.2046.$$

On the other hand, since $-180° \le \theta \le 180°$, $-90° \le \frac{\theta}{2} \le 90°$,

$\therefore \frac{\theta}{2} \simeq 48.43°, -11.56°$ and $\theta \simeq 96.9°, -23.1°$.

Exercise. Solve by the two methods: $2\sin\theta + 3\cos\theta = -1$ for $\theta : -180° \le \theta \le 180°$.

<u>Ans:</u> $-72.4°, 139.8°$.   ◇

Exercise. Solve by the two methods: $5\cos\theta - 3\sin\theta = 1$ for $\theta : -180° \le \theta \le 180°$.

<u>Ans:</u> $49.1°, -101.1°$.   ◇

Chapter 17

Ptolemy's Theorem

Remark 17.1. The theorem of Ptolemy is of immense historical importance, for it has been shown that the whole of Plane Trigonometry is deducible from this one result. Ptolemy of Alexandria (c. 90 AD–c. 180 AD) was an astronomer. His famous work, the Almagest, in several volumes, was translated into Arabic c. 817 AD.

17.1 Ptolemy's theorem [Claude Ptolemy, c. 139 AD]

Theorem 17.1 (Ptolemy's Theorem).
In a cyclic quadrilateral ABCD,

$$AB \cdot CD + AD \cdot CB = AC \cdot BD.$$

— *i.e. the sum of the products of opposite sides of a cyclic quadrilateral is equal to the product of its diagonals.*

Trigonometric proof. Refer to Fig. 17.1.1.

R = radius of circumscribing circle. Let the sides subtend angles $2x$, $2y$, $2z$, $2w$ at centre O.

$x + y + z + w = 180°$.

The 4 triangles with vertices at O, shown, are all isosceles. So we have $AB = 2R\sin x$, $BC = 2R\sin y$, $CD = 2R\sin z$ and $DA = 2R\sin w$.

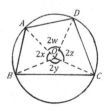

Fig. 17.1.1

73

Disregarding the factor $2R$, we must show

$$(*) \quad \sin x \sin z + \sin y \sin w = \sin(x + y) \sin(x + w).$$

Thus LHS of $(*) = \sin x \sin(x + w + y) + \sin y \sin w$, on eliminating z,

$$= \sin x \sin(x + w + y) + \sin y \sin(x + w - x), \text{clearly},$$

$$= \sin x \big\{ \sin(x + w) \cos y + \cos(x + w) \sin y \big\} +$$

$$\sin y \big\{ \sin(x + w) \cos x - \cos(x + w) \sin x \big\},$$

using addition formulae,

$$= \sin(x + w)(\sin x \cos y + \cos x \sin y),$$

on cancelling out two of the terms,

$$= \sin(x + y) \sin(x + w)$$

$$= \text{RHS of } (*).$$

$\square$

Geometric proof. Refer to Fig. 17.1.2.

X is chosen on the diagonal BD so that $D\hat{A}X = B\hat{A}C$.

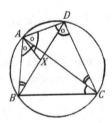

Since $A\hat{D}X = A\hat{C}B$, $\triangle ADX$ is similar to $\triangle BCA$; so $\frac{DX}{BC} = \frac{AD}{AC}$, and

$$DX \cdot AC = AD \cdot BC. \qquad (17.1)$$

Again $B\hat{A}X = D\hat{A}C$, and since $A\hat{B}X = A\hat{C}D$, $\triangle ABX$ is similar to $\triangle ACD$; so $\frac{BX}{CD} = \frac{AB}{AC}$, and

Fig. 17.1.2

$$BX \cdot AC = AB \cdot DC. \qquad (17.2)$$

Adding (17.1) to (17.2) produces

$$AC(DX + BX) = AD \cdot BC + AB \cdot DC,$$

$$\therefore AC \cdot BD = AB \cdot DC + AD \cdot BC.$$

$\square$

Chapter 18

Morley's Trisector Theorem

18.1 Morley's theorem [Frank Morley, 1860–1937]

Theorem 18.1 (Morley's Trisector Theorem).
*In any triangle ABC the three points
of intersection, L, M, N of adjacent pairs of angle trisectors always
form an equilateral triangle. (Refer
to Fig. 18.1.1.)*

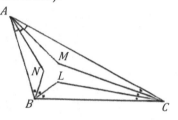

Note: Proved in 1899, this result is
also known as Morley's Miracle. For
a purely geometric proof, see p. 24 of
reference [8].

Fig. 18.1.1

Trigonometric proof. We prove

(i) $4\sin\frac{C}{3}\sin\left(60° + \frac{C}{3}\right)\sin\left(60° - \frac{C}{3}\right) = \sin C$:

$$LHS = 2\sin\frac{C}{3}\cos\frac{2C}{3} - \cos 120°$$

$$= \sin\frac{C}{3}\left\{2\left(1 - 2\sin^2\frac{C}{3}\right) + 1\right\}$$

$$= 3\sin\frac{C}{3} - 4\sin^3\frac{C}{3}$$

$$= \sin C, \text{ from problem 1) in 13.2.}$$

(ii) From $\triangle ANB$, $\dfrac{AN}{\sin \frac{B}{3}} = \dfrac{c}{\sin \left(180° - \frac{A+B}{3}\right)}$.

$$= 2R\dfrac{\sin C}{\sin \frac{A+B}{3}}$$

$$= 2R\dfrac{\sin C}{\sin \left(60° - \frac{C}{3}\right)}$$

$$= 8R\sin \dfrac{C}{3}\sin \left(60° + \dfrac{C}{3}\right), \text{ after using (i)}$$

$\therefore AN = 8R\sin \frac{B}{3}\sin \frac{C}{3}\sin \left(60° + \frac{C}{3}\right)$;

and similarly $AM = 8R\sin \frac{C}{3}\sin \frac{B}{3}\sin \left(60° + \frac{B}{3}\right)$.

(iii) $MN^2 = AN^2 + AM^2 - 2AN \cdot AM\cos \frac{A}{3}$. For convenience, put $8R\sin \dfrac{B}{3}\sin \dfrac{C}{3} = K$.

$$\dfrac{MN^2}{K^2} = \sin^2\left(60° + \dfrac{C}{3}\right) + \sin^2\left(60° + \dfrac{B}{3}\right)$$

$$- 2\sin \left(60° + \dfrac{C}{3}\right)\sin \left(60° + \dfrac{B}{3}\right)\cos \dfrac{A}{3}$$

$$= \dfrac{1}{2}\left\{1 - \cos \left(120° + \dfrac{2C}{3}\right)\right\} + \dfrac{1}{2}\left\{1 - \cos \left(120° + \dfrac{2B}{3}\right)\right\}$$

$$- \cos \dfrac{A}{3}\left\{\cos \dfrac{B-C}{3} - \cos \left(120° + \dfrac{B+C}{3}\right)\right\}.$$

The first two cosine terms $= 1 - 2\cos \left(120° + \frac{B+C}{3}\right)\cos \frac{B-C}{3}$, and $\cos \left(120° + \frac{B+C}{3}\right) = -\cos \frac{A}{3}$.

$\therefore \frac{MN^2}{K^2} = 1 + \cos \frac{A}{3}\cos \frac{B-C}{3} - \cos \frac{A}{3}\left(\cos \frac{B-C}{3} + \cos \frac{A}{3}\right) = \sin^2 \frac{A}{3}$.

$\therefore MN = K\sin \frac{A}{3} \equiv 8R\sin \frac{A}{3}\sin \frac{B}{3}\sin \frac{C}{3}$.

Symmetry shows $MN = ML = LN$.

$\square$

Exercise. If $\triangle ABC$ is itself equilateral, show that $\dfrac{\text{area} \triangle ABC}{\text{area} \triangle LMN} \simeq 29.3$.

Hint: Take the square of the ratio of their sides. $\diamond$

Chapter 19

Cyclic Quadrilaterals and Brahmagupta's Formula

19.1 Notation

Refer to Fig. 19.1.1. $ABCD$ is a convex quadrilateral with sides $AB = a$, $BC = b$, $CD = c$, $DA = d$. The diagonals AC, BD meet at P with $A\hat{P}B = \theta$.

Let $AC = x$, $BD = y$. Put $\sigma = \frac{1}{2}(a+b+c+d)$. It is the semi-perimeter.

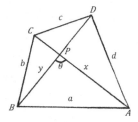

Put $S = \text{area } ABCD$. Note that

Fig. 19.1.1

$$S = \text{area } \triangle ABD + \text{area } \triangle CBD = \frac{1}{2}y(AP\sin\theta) + \frac{1}{2}y(CP\sin\theta)$$

$$= \frac{1}{2}y(AP + CP)\sin\theta = \frac{1}{2}xy\sin\theta.$$

19.2 Theorem

Theorem 19.1. *In a cyclic quadrilateral, with notation as above, we have*

(i) $\sin B = \frac{2S}{ab+cd}(=\sin D)$; $\quad \cos B = \frac{a^2+b^2-c^2-d^2}{2(ab+cd)}(=-\cos D)$;

(ii) $S = \sqrt{(\sigma-a)(\sigma-b)(\sigma-c)(\sigma-d)}$;

— Brahmagupta's formula (c. 630 AD)

(iii) $x = \sqrt{\frac{(ac+bd)(ad+bc)}{ab+cd}}$, $\quad y = \sqrt{\frac{(ac+bd)(ab+cd)}{ad+bc}}$;

77

(iv) *the circumradius* $= \frac{1}{4S}\sqrt{(ab+cd)(ac+bd)(ad+bc)}$.

Proof. (i) $S = $ area $\triangle ABC + $ area $\triangle ADC$,

$\therefore 2S = ab\sin B + cd\sin D = (ab+cd)\sin B \dots\dots\dots\dots (*)$, giving $\sin B$;

$a^2 + b^2 - 2ab\cos B = x^2 = c^2 + d^2 - 2cd\cos D = c^2 + d^2 + 2cd\cos B$,

$\therefore a^2 + b^2 - c^2 - d^2 = 2(ab+cd)\cos B \dots\dots\dots\dots (**)$, giving $\cos B$;

(ii) We eliminate B from $(*)$ and $(**)$:
$$4(ab+cd)^2 = (4S)^2 + (a^2+b^2-c^2-d^2)^2.$$

$\therefore 16S^2 = \{2(ab+cd) + (a^2+b^2-c^2-d^2)\}\times$
$$\{2(ab+cd) - (a^2+b^2-c^2-d^2)\}$$
(on factorising the difference of squares)
$$= \{(a+b)^2 - (c-d)^2\}\{(c+d)^2 - (a-b)^2\}$$
$$= (a+b+c-d)(a+b-c+d)(a-b+c+d)(-a+b+c+d)$$
(the complete factorisation)
$$= (2\sigma - 2d)(2\sigma - 2c)(2\sigma - 2b)(2\sigma - 2a),$$
(introducing the semi-perimeter σ)

$\therefore S^2 = (\sigma - a)(\sigma - b)(\sigma - c)(\sigma - d).$

(iii) From (i) we may eliminate $\cos B$:
$$2\cos B = \frac{a^2+b^2-x^2}{ab} = \frac{x^2-c^2-d^2}{cd}.$$

$\therefore (ab+cd)x^2 = cd(a^2+b^2) + ab(c^2+d^2)$
$$= (ac+bd)(ad+bc).$$

Thus x is obtained; y is found similarly.

(iv) The circumradius is
$$\frac{1}{2}\frac{x}{\sin B} = \sqrt{\frac{(ac+bd)(ad+bc)}{ab+cd}} \Big/ \frac{4S}{ab+cd} \quad \text{(from (iii) and (i))}$$
$$= \frac{1}{4S}\sqrt{(ab+cd)(ac+bd)(ad+bc)}.$$

$\square$

19.3   Problems / exercises

1) Express $\tan \frac{B}{2}$ in terms of the sides of the cyclic quadrilateral.

Solution.

$$\tan^2 \frac{B}{2} = \frac{1 - \cos B}{1 + \cos B}$$

$$= \frac{2(ab + cd) - (a^2 + b^2 - c^2 - d^2)}{2(ab + cd) + (a^2 + b^2 - c^2 - d^2)} \quad \text{(from (i) of Theorem 19.1)}$$

$$= \frac{(c + d)^2 - (a - b)^2}{(a + b)^2 - (c - d)^2}.$$

Exercise.

(i) Use the values of x, y in the Theorem 19.1 part (iii) to re-prove Ptolemy's Theorem (i.e., Theorem 17.1), viz. $xy = ac + bd$.

(ii) A cyclic quadrilateral has $a = 3.5\,\text{cm}$, $b = 4.0\,\text{cm}$, $c = 4.5\,\text{cm}$, $d = 5.0\,\text{cm}$.

Find, to no more that 3 s.f., σ, S, x, $\hat{B}$ (from $\cos B$), and the circumradius.

Draw the quadrilateral from this information.

<u>Ans</u>: $8.5\,\text{cm}, 17.7\,\text{cm}^2, 5.90\,\text{cm}, 103°, 3.04\,\text{cm}$.

$\diamond$

Chapter 20

Graphs of the Six Trigonometrical Ratios

20.1 Notes on the graphs

The six ratios $\sin\theta, \cdots, \cot\theta$ are graphed below for θ from $0°$ to $180°$. It is easy to extend these graphs to other domains, after noting the following:

(i) $\sin\theta, \tan\theta, \operatorname{cosec}\theta, \cot\theta$ are odd functions: e.g. $\sin(-\theta) = -\sin\theta, \cdots,$ $\cot(-\theta) = -\cot\theta$; and $\cos\theta, \sec\theta$ are even functions: $\cos(-\theta) = \cos\theta,$ $\sec(-\theta) = \sec\theta.$

(ii) All are periodic functions: $\sin(\theta \pm 360°) = \sin\theta, \cdots, \cot(\theta \pm 360°) =$ $\cot\theta$. $360°$ is a period. $\tan\theta, \cot\theta$ have shorter periods: $180°$.

(iii) $\sin\theta, \cos\theta$ are related by $90°$ shifts in θ; e.g. $\sin(\theta + 90°) = \cos\theta.$ Also $\operatorname{cosec}\theta, \sec\theta$ are similarly related; e.g., $\operatorname{cosec}(\theta + 90°) = \sec\theta.$

(iv) $|\sin\theta| \leq 1$, $|\cos\theta| \leq 1$ for all θ and $|\operatorname{cosec}\theta| \geq 1$, $|\sec\theta| \geq 1$; $\tan\theta,$ $\cot\theta$ can take all values.

20.2 Graphs of $\sin\theta, \cos\theta, \tan\theta$

Refer to Figs. 20.2.1, 20.2.2, and 20.2.3.

$\sin\theta$ and $\cos\theta$ are wave-like in shape; $\tan\theta$ has branches and an asymptote at $\theta = 90°$.

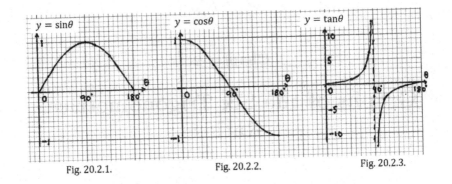

Fig. 20.2.1.   Fig. 20.2.2.   Fig. 20.2.3.

20.3   Graphs of cosec θ, sec θ, cot θ

Refer to Figs. 20.3.1, 20.3.2, and 20.3.3.

cosec θ and cot θ have asymptotes at $\theta = 0°, 180°$; sec θ has an asymptote at $\theta = 90°$.

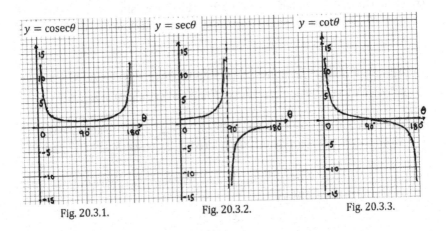

Fig. 20.3.1.   Fig. 20.3.2.   Fig. 20.3.3.

20.4   Problems / exercises

Sketch the graph $y = 1 + 2\cos(3\theta - 15°)$ over the domain $-90° \le \theta \le 90°$.

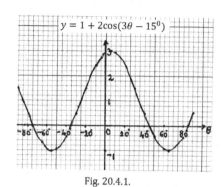

Fig. 20.4.1.

Solution. Refer to Fig. 20.4.1.

The graph should be symmetrical about the value given by $3\theta - 15° = 0$, i.e. $\theta = 5°$ (a vertical line). y must have a maximum value of $1 + 2 = 3$ (when $\theta = 5°$) and a minimum value of $1 - 2 = -1$ (when $3\theta - 15° = \pm180°$, i.e. $\theta = 65°$, $-55°$). Zero values of y occur where $3\theta - 15° = 120°$, $-120°$, i.e. $\theta = 45°$, $-35°$, $85°$, $-75°$.

The sketch follows from this information: there is no need for a point by point plot.

Exercise. Solve the equation $1 + 2\cos(3\theta - 15°) = 2$, for θ in the range $-90°$ to $90°$.

<u>Ans:</u> $\theta = -15°, 25°$.

◊

Chapter 21

Graphs of the Six Inverse Trigonometrical Ratios

21.1 Notes on the graphs

If, for instance, we are given $\sin \theta = 0.5$, then if θ is an angle of a triangle, $\theta = 30°$ or $150°$. But there are other solutions for θ (e.g. $-210°$) when the domain for θ is extended. The principal value here is written $\sin^{-1}\left(\frac{1}{2}\right) = 30°$, which is the value given by a calculator ($\frac{\pi}{6}$ when radian mode is chosen): calculators always display the smallest of numerical possibilities.

Other examples:

$\cos^{-1}\left(-\frac{1}{\sqrt{2}}\right) = 135°$, rather than $-225°$ or $-135°$; $\tan^{-1}(-1) = -45°$, rather than $135°$.

21.2 Graphs of $\sin^{-1} x, \cos^{-1} x, \tan^{-1} x$

Refer to Figs. 21.2.1, 21.2.2 and 21.2.3.

21.3 Graphs of $\operatorname{cosec}^{-1} x, \sec^{-1} x, \cot^{-1} x$

Refer to Figs. 21.3.1, 21.3.2 and 21.3.3.

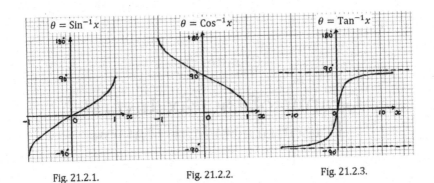

Fig. 21.2.1.   Fig. 21.2.2.   Fig. 21.2.3.

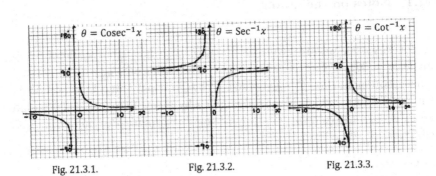

Fig. 21.3.1.   Fig. 21.3.2.   Fig. 21.3.3.

21.4   Problems / exercises

1) Find $\sin^{-1} \tan \sec^{-1}(-\sqrt{2})$.

Solution. If $\sec^{-1}(-\sqrt{2}) = \theta$, then $\sec\theta = -\sqrt{2}$, $\therefore \cos\theta = -\frac{1}{\sqrt{2}}$, so that $\theta = 135°$, the principal value of θ.

$\tan\theta = -1$, and finally $\sin^{-1}(-1) = -90°$, the principal value. In conclusion $\sin^{-1} \tan \sec^{-1}(-\sqrt{2}) = -90°$ $(= -\frac{\pi}{2}$ rads).

Exercise. Show that $\frac{1}{2}\cos^{-1}(1 - 2x^2) = \sin^{-1}|x|$ $(-1 \leq x \leq 1)$. $\diamond$

Chapter 22

Addition Formulae for Inverse Functions & Rutherford's and Machin's Formulae

22.1 Theorem

Theorem 22.1. $\tan^{-1} x \pm \tan^{-1} y = \tan^{-1}\left(\dfrac{x \pm y}{1 \mp xy}\right).$

Proof. Put $\tan^{-1}x = \alpha$, $\tan^{-1}y = \beta$, so $x = \tan\alpha$, $y = \tan\beta$.
Then

$$\tan(\alpha \pm \beta) = \frac{\tan\alpha \pm \tan\beta}{1 \mp \tan\alpha\tan\beta}, \text{ from Theorem 14.1,}$$
$$= \frac{x \pm y}{1 \mp xy},$$

so, $\alpha \pm \beta = \tan^{-1}x \pm \tan^{-1}y = \tan^{-1}\left(\frac{x\pm y}{1\mp xy}\right).$ □

Corollary 22.1. *The special case* $y = x$ *gives*

$$2\tan^{-1} x = \tan^{-1}\left(\frac{2x}{1 - x^2}\right).$$

22.2 Rutherford's formula [Ernest Rutherford, 1871–1937, N.Z. Physicist]

Theorem 22.2 (Rutherford's Formula).

$$4\tan^{-1}\left(\frac{1}{5}\right) - \tan^{-1}\left(\frac{1}{70}\right) + \tan^{-1}\left(\frac{1}{99}\right) = 45°. \quad \left(= \frac{\pi}{4}\text{ rads}\right)$$

Proof. We apply Corollary 22.1 twice,

$$2\tan^{-1}\left(\frac{1}{5}\right) = \tan^{-1}\left(\frac{\frac{2}{5}}{1-\frac{1}{25}}\right)$$

$$= \tan^{-1}\left(\frac{5}{12}\right),$$

so that

$$4\tan^{-1}\left(\frac{1}{5}\right) = \tan^{-1}\left(\frac{\frac{5}{6}}{1-\frac{25}{144}}\right)$$

$$= \tan^{-1}\left(\frac{120}{119}\right).$$

Next, by Theorem 22.1,

$$\tan^{-1}\left(\frac{1}{70}\right) - \tan^{-1}\left(\frac{1}{99}\right) = \tan^{-1}\left(\frac{\frac{1}{70}-\frac{1}{99}}{1+\frac{1}{70}\frac{1}{99}}\right)$$

$$= \tan^{-1}\left(\frac{29}{6931}\right)$$

$$= \tan^{-1}\left(\frac{1}{239}\right),$$

so the three-term sum $= \tan^{-1}\left(\dfrac{120}{119}\right) - \tan^{-1}\left(\dfrac{1}{239}\right)$

$$= \tan^{-1}\left(\frac{\frac{120}{119}-\frac{1}{239}}{1+\frac{120}{119}\frac{1}{239}}\right)$$

$$= \tan^{-1}\left(\frac{28561}{28561}\right) = 45°, \quad \text{i.e., } \frac{\pi}{4}\text{ rads}.$$

□

22.3 Machin's formula

Theorem 22.3 (Machin's Formula).

$$4\tan^{-1}\left(\frac{1}{5}\right) - \tan^{-1}\left(\frac{1}{239}\right) = 45°, i.e., \frac{\pi}{4}\text{ rads}.$$

Proof. LHS $= 4\tan^{-1}\left(\frac{1}{5}\right) - \left\{\tan^{-1}\left(\frac{1}{70}\right) - \tan^{-1}\left(\frac{1}{99}\right)\right\}$, from the calculation in the proof of Rutherford's formula. Hence LHS $= 45°$.   □

Note: Rutherford's result is of historical importance; it provides a way of calculating π. From courses in calculus it is known that

$$\tan^{-1}x = x - \frac{x^3}{3} + \frac{x^5}{5} - \frac{x^7}{7} + \cdots ,$$

a convergent series valid for $-1 < x < 1$; where $\tan^{-1}x$ is in radians (between $-\frac{\pi}{4}$ and $\frac{\pi}{4}$). The LHS of Rutherford's formula is therefore

$$4\left\{ \frac{1}{5} - \frac{(\frac{1}{5})^3}{3} + \frac{(\frac{1}{5})^5}{5} - \cdots \right\} - \left\{ \frac{1}{70} - \frac{(\frac{1}{70})^3}{3} + \frac{(\frac{1}{70})^5}{5} - \cdots \right\} \\ + \left\{ \frac{1}{99} - \frac{(\frac{1}{99})^3}{3} + \frac{(\frac{1}{99})^5}{5} - \cdots \right\}. \tag{22.1}$$

22.4   Problems / exercises

1) Using the indicated 9 terms of (22.1) at the end of 22.3, find $\frac{\pi}{4}$ to the extent that a calculator allows.

Solution. The total from the 9 terms comes to 0.7853979431, while $\frac{\pi}{4} = $ 0.7853981634.

Exercise. Using only the corresponding 9 terms in Machin's formula (Theorem 22.3), find $\frac{\pi}{4}$.

<u>Ans:</u> 0.7854052573.   ◇

2) Solve for: $\tan^{-1}\left(x + \frac{2}{x}\right) - \tan^{-1}\left(x - \frac{2}{x}\right) = \tan^{-1}\left(\frac{4}{x}\right)$.

Solution. By the theorem,

$$\text{LHS} = \tan^{-1}\left(\frac{\left(x + \frac{2}{x}\right) - \left(x - \frac{2}{x}\right)}{1 + \left(x^2 - \frac{4}{x^2}\right)} \right)$$

$$= \tan^{-1}\left(\frac{\frac{4}{x}}{1 + \left(x^2 - \frac{4}{x^2}\right)} \right)$$

$$= \text{RHS}$$

$$= \tan^{-1}\left(\frac{4}{x}\right),$$

provided $1 + \left(x^2 - \frac{4}{x^2}\right) = 1$, $\therefore x^4 = 4$, $x^2 = 2$, $x = \pm\sqrt{2}$.

Exercise. Prove that $3\tan^{-1}\left(\dfrac{1}{4}\right) - \tan^{-1}\left(\dfrac{1}{20}\right) + \tan^{-1}\left(\dfrac{1}{1985}\right) = \dfrac{\pi}{4}$.

Prove (i) $\sin^{-1} x \pm \sin^{-1} y = \sin^{-1}\left\{x\sqrt{1-y^2} \pm y\sqrt{1-x^2}\right\}$.

(ii) $\cos^{-1} x \pm \cos^{-1} y = \cos^{-1}\left\{xy \mp \sqrt{(1-x^2)(1-y^2)}\right\}$, both rarely used because of signs.   ◇

Chapter 23

Solving Simultaneous Equations

Remark In this chapter we confine ourselves to solving two equations in two unknown angles only. As there are no general methods for solving them, they would need to be considered on their own merits.

23.1 Problems / exercises

1) Solve $\begin{cases} 33\cos\theta = 52\cos\varphi; \\ 8\sin\theta + 13\sin\varphi = 16; \end{cases}$ for positive angles less than $90°$.

Solution. Multiply the second equation by 4, and then eliminate φ using the first equation:

$$33^2\cos^2\theta = 52^2 - 52^2\sin^2\varphi$$
$$= 2704 - \{4(16 - 8\sin\theta)\}^2$$
$$= 2704 - 1024(2 - \sin\theta)^2,$$

$1089(1 - \sin^2\theta) = 2704 - 1024(4 - 4\sin\theta + \sin^2\theta)$, a quadratic in $\sin\theta$.

$\therefore 65\sin^2\theta + 4096\sin\theta - 2481 = 0$,

$(5\sin\theta - 3)(13\sin\theta + 827) = 0$,

$\sin\theta = \frac{3}{5}$ only.

From the second equation, $\sin\varphi = \frac{1}{13}\left(16 - \frac{24}{5}\right) = \frac{56}{65}$.

Hence, $\theta \simeq 36.9°$, $\varphi \simeq 59.5°$. (acute angles angles as required)

2) Given $\begin{cases} \tan\theta = 9\tan\varphi; \\ 5\cos(\theta - \varphi) - 4\cos(\theta + \varphi) = 7; \end{cases}$ prove that $\sin^2\varphi = 0.6$.

Hence find θ, φ as acute angles.

Solution. From the second equation by Theorem 13.1 we have,

$5(\cos\theta\cos\varphi + \sin\theta\sin\varphi) - 4(\cos\theta\cos\varphi - \sin\theta\sin\varphi) = 7,$

i.e. $\cos\theta\cos\varphi + 9\sin\theta\sin\varphi = 7$, it follows that

$1 + 9\tan\theta\tan\varphi = 7\sec\theta\sec\varphi.$

Using the first equation, we find $1 + \tan^2\theta \equiv \sec^2\theta = 7\sec\theta\sec\varphi$.

$\therefore \sec\theta = 7\sec\varphi.$

Squaring, $1 + 81\tan^2\varphi = 49\sec^2\varphi,$

$\cos^2\varphi + 81\sin^2\varphi = 49,$

$80\sin^2\varphi = 48,$

$\sin^2\varphi = 0.6.$

Thus $\varphi \simeq 50.77°$, making $\tan\theta \simeq 11.023, \theta \simeq 34.82°$.

Exercise. Solve $\begin{cases} \cos\theta + 5\cos\varphi = 5; \\ \sin\theta = 2\sin\varphi; \end{cases}$ for angles less than $90°$.

<u>Ans:</u> $\theta \simeq 60.3°, \varphi \simeq 25.7°$.   ◇

Chapter 24

The Problem of Elimination

24.1 Introduction

The process of removing a variable angle from two given equations involving functions of the angle and coordinates x, y is called elimination. To take a simple case: from the equations $x = a \sec\theta$, $y = b \tan\theta$, we easily find that the eliminant is $\frac{x^2}{a^2} - \frac{y^2}{b^2} = 1$, a hyperbolic curve. Indeed, $x = a \sec\theta$, $y = b \tan\theta$ are possible parametric equations for a standard hyperbola.

24.2 Problems / exercises

1) Prove that the eliminant of θ from the pair of equations

$$x \sin\theta = 2\cos^3\theta, \quad y = \sin^2\theta$$

is $x^2 y = 4(1 - y)^3$.

Solution.

$$x^2 y = \left(\frac{2\cos^3\theta}{\sin\theta} \right)^2 \sin^2\theta$$
$$= 4\cos^6\theta$$
$$= 4(1 - \sin^2\theta)^3 = 4(1 - y)^3.$$

2) If $\cot\theta(1 + \sin\theta) = 4a$ and $\cot\theta(1 - \sin\theta) = 4b$, find the eliminant of θ.

Solution. We obtain: $\cot\theta + \cos\theta = 4a$, $\cot\theta - \cos\theta = 4b$,

$\therefore \cos \theta = 2(a - b)$ and $\cot \theta = 2(a + b)$.

By division, $\sin \theta = \frac{a-b}{a+b}$.

The relation $\cos^2\theta + \sin^2 \theta = 1$ gives

$$\{2(a - b)\}^2 + \left\{\frac{a - b}{a + b}\right\}^2 = 1,$$

$$\therefore 4(a^2 - b^2)^2 + (a - b)^2 = (a + b)^2,$$

reducing to $(a^2 - b^2)^2 = ab$.

Exercise. Eliminate φ between $x(1 - \cos \varphi) = a + y \sin \varphi$ and $x \sin \varphi + a = y(1 - \cos \varphi)$; also between $x \cos \varphi + y \sin \varphi = p$ and $-x \sin \varphi + y \cos \varphi = 0$.

<u>Ans:</u> $x + y = a$ (a line); $x^2 + y^2 = p^2$ (a circle).   ◇

3) Eliminate θ between the equations

$$\begin{cases} ax \sec \theta - by \operatorname{cosec} \theta = a^2 - b^2, \\ ax \sec \theta \tan \theta + by \operatorname{cosec} \theta \cot \theta = 0. \end{cases}$$

[The eliminant here gives the locus of the centres of curvature for an ellipse $\frac{x^2}{a^2} + \frac{y^2}{b^2} = 1$; refer to a book on coordinate geometry.]

Solution. On eliminating y from the equations we find

$$ax \sec \theta[\cot \theta + \tan \theta] = (a^2 - b^2) \cot \theta,$$

$$\therefore ax \sec \theta = \frac{(a^2 - b^2) \cot \theta}{\cot \theta + \tan \theta} = (a^2 - b^2) \cos^2 \theta.$$

On eliminating x from the same pair of equations, we find

$$by \operatorname{cosec} \theta[\cot \theta + \tan \theta] = -(a^2 - b^2) \tan \theta,$$

$$\therefore by \operatorname{cosec} \theta = -\frac{(a^2 - b^2) \tan \theta}{\cot \theta + \tan \theta} = -(a^2 - b^2) \sin^2 \theta.$$

Thus,

$$ax = (a^2 - b^2) \cos^3 \theta, \quad by = -(a^2 - b^2)\sin^3\theta,$$

$$\therefore \left(\frac{ax}{a^2 - b^2}\right)^{\frac{2}{3}} + \left(\frac{-by}{a^2 - b^2}\right)^{\frac{2}{3}} = \cos^2 \theta + \sin^2 \theta = 1,$$

or,

$$(ax)^{\frac{2}{3}} + (by)^{\frac{2}{3}} = (a^2 - b^2)^{\frac{2}{3}},$$

representing an astroid. [Assume $a > b$]

Exercise. Give a sketch of the astroid in the problem above.   ◇

Chapter 25

Angle Bisectors and the Incentre, I

25.1   Incircle; centre I, radius r

Refer to Fig. 25.1.1.

The incircle of $\triangle ABC$ has its centre at I. I is at the intersection of the bisectors of $\hat{A}$, $\hat{B}$, $\hat{C}$. r is the radius of the incircle. X, Y, Z are the contact points with the sides.

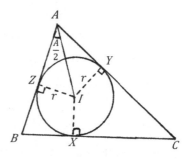

Fig. 25.1.1

25.2   Theorem

Theorem 25.1.

(i) $AY = AZ = s - a$;

(ii) $r = \dfrac{\Delta}{s} = (s - a)\tan\dfrac{A}{2} = 4R\sin\dfrac{A}{2}\sin\dfrac{B}{2}\sin\dfrac{C}{2}$;

(iii) $AI = r\,\mathrm{cosec}\,\dfrac{A}{2} = (s - a)\sec\dfrac{A}{2} = 4R\sin\dfrac{B}{2}\sin\dfrac{C}{2}$.

Proof. (i) $AY = AZ$, being equal tangents to the incircle from A.

Put $AY = AZ = x$; also put $BX = BZ = y$, $CX = CY = z$,

$2x + 2y + 2z = $ perimeter of $\triangle ABC = a + b + c = 2s$,

so that $a = BX + XC = y + z = s - x$,

95

$\therefore AY = AZ = s - a.$

(ii) Since area $\triangle ABC$ = area $\triangle BIC$ + area $\triangle CIA$ + area $\triangle AIB$,

$\Delta = \frac{1}{2}ar + \frac{1}{2}br + \frac{1}{2}cr = rs,$

$\therefore r = \frac{\Delta}{s}$, the simplest formula for r.

Also,

$$4R\sin\frac{A}{2}\sin\frac{B}{2}\sin\frac{C}{2} = 4R\sqrt{\frac{(s-b)(s-c)}{bc}}\sqrt{\frac{(s-c)(s-a)}{ca}}\sqrt{\frac{(s-a)(s-b)}{ab}},$$

from Theorem 15.1,

$$= 4R\frac{(s-a)(s-b)(s-c)}{abc}, \text{ on reduction,}$$

$$= \frac{\frac{\Delta^2}{s}}{\Delta}, \text{ from Hero's formula and Theorem 7.1,}$$

$$= \frac{\Delta}{s}$$

$$= r.$$

(iii) From $\triangle AIZ$ again,

$$AI = r\operatorname{cosec}\frac{A}{2} = (s-a)\sec\frac{A}{2} \text{ (are seen immediately)}$$

$$= 4R\sin\frac{B}{2}\sin\frac{C}{2}, \text{ from the last part in (ii).}$$

$\square$

Chapter 26

Altitudes, the Orthocentre, H and the Pedal Triangle

26.1 Orthocentre, H

The altitudes of $\triangle ABC$ intersect at a point H. H is the orthocentre. If AD, BE, CF are the altitudes, $\triangle DEF$ is the orthic triangle of $\triangle ABC$ and is also known as the pedal triangle. The word 'pedal' triangle is a more general concept; the orthic triangle is a special case.

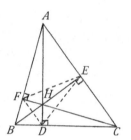

Fig. 26.1.1

26.2 Theorem

Theorem 26.1. *Assuming $\triangle ABC$ is acute-angled, H is inside the triangle, and*

(i) $AH = 2R\cos A = \sqrt{4R^2 - a^2}, \quad DH = 2R\cos B \cos C$;

(ii) *DH bisects $E\hat{D}F$, H is the incentre of $\triangle DEF$ and $E\hat{D}F = 180° - 2A$;*

(iii) $EF = a\cos A = R\sin(2A)$;

(iv) $\text{area} \triangle DEF = \dfrac{1}{2}R^2 \sin(2A)\sin(2B)\sin(2C)$;

(v) *The circumradius of $\triangle DEF = \dfrac{1}{2}R$.*

Proof. Refer to Fig. 26.1.1

(i) From $\triangle HAE$, $AH = \dfrac{AE}{\cos H\hat{A}E}$

$\qquad\qquad = \dfrac{c\cos A}{\sin C}$

$\qquad\qquad = 2R\cos A$

$\qquad\qquad = \sqrt{4R^2 - (2R\sin A)^2}$

$\qquad\qquad = \sqrt{4R^2 - a^2}.$

Also, $DH = BD\tan H\hat{B}D$

$\qquad\quad = c\cos B\cot C$

$\qquad\quad = 2R\sin C\cos B\,\dfrac{\cos C}{\sin C}$

$\qquad\quad = 2R\cos B\cos C.$

(ii) Quadrilaterals $HDBF$, $HDCE$ are cyclic, so $H\hat{D}F = H\hat{B}F = 90° - A$; also $H\hat{D}E = H\hat{C}E = 90° - B$. Thus, HD bisects $E\hat{D}F$, therefore $E\hat{D}F = 180° - 2A$. Similarly HE bisects $D\hat{E}F$, HF bisects $E\hat{F}D$, making H the incentre of $\triangle DEF$.

(iii) AH is the diameter of the cyclic quadrilateral $AEHF$, $\frac{EF}{\sin A} = AH$,

$\qquad \therefore EF = 2R\cos A\sin A = a\cos A = R\sin(2A).$

(iv) area $\triangle DEF = \dfrac{1}{2}DE \cdot DF\sin E\hat{D}F$

$\qquad\qquad\qquad = \dfrac{1}{2}R\sin(2C)R\sin(2B)\sin(180° - 2A)$, from part (iii),

$\qquad\qquad\qquad = \dfrac{1}{2}R^2\sin(2A)\sin(2B)\sin(2C).$

(v) Circumradius of $\triangle DEF = \dfrac{1}{2}\dfrac{EF}{\sin E\hat{D}F} = \dfrac{1}{2}\dfrac{R\sin(2A)}{\sin(2A)} = \dfrac{1}{2}R.$
[Confirmed by Theorem 28.1 (iv).]   □

Exercise. If A is obtuse, H lies outside the triangle. Draw the alternate figure, and show that the sides of the pedal triangle DEF are $-a\cos A$, $b\cos B$, $c\cos C$; and that the angles are $2A - 180°$, $2B$, $2C$.   ◇

Chapter 27

The Distances OI, OH, IH

27.1 Theorem

Theorem 27.1. *In* $\triangle ABC$,

(i) AI *bisects* $O\hat{A}H$;

(ii) $OI^2 = R^2\left(1 - 8\sin\dfrac{A}{2}\sin\dfrac{B}{2}\sin\dfrac{C}{2}\right) = R^2 - 2Rr$;

(iii) $OH^2 = R^2(1 - 8\cos A\cos B\cos C)$;

(iv) $IH^2 = 2r^2 - 4R^2\cos A\cos B\cos C$.

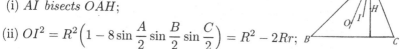

Fig. 27.1.1

Proof. Refer to Fig. 27.1.1, which is drawn with $C > B$.

(i) $B\hat{O}A = 2C$, therefore $B\hat{A}O = 90° - C$,

$$O\hat{A}I = \frac{1}{2}A - (90° - C) = \frac{1}{2}A - \left(\frac{A+B+C}{2} - C\right) = \frac{C-B}{2},$$

$$I\hat{A}H = \frac{1}{2}A - H\hat{A}C = \frac{1}{2}A - (90° - C) = \frac{C-B}{2}.$$

Hence AI bisects $O\hat{A}H$.

(ii) $OI^2 = OA^2 + AI^2 - 2OA \cdot AI\cos O\hat{A}I$

$$= R^2 + \left(4R\sin\frac{B}{2}\sin\frac{C}{2}\right)^2 - 2R\left(4R\sin\frac{B}{2}\sin\frac{C}{2}\right)\cos\frac{C-B}{2},$$

on using Theorem 25.1 (iii) and Theorem 27.1 (i) above,

99

$$= R^2 \left[1 + 8 \sin \frac{B}{2} \sin \frac{C}{2} \left\{ 2 \sin \frac{B}{2} \sin \frac{C}{2} \right. \right.$$

$$\left. \left. - \left(\cos \frac{C}{2} \cos \frac{B}{2} + \sin \frac{C}{2} \sin \frac{B}{2} \right) \right\} \right]$$

$$= R^2 \left[1 + 8 \sin \frac{B}{2} \sin \frac{C}{2} \left\{ \sin \frac{B}{2} \sin \frac{C}{2} - \cos \frac{B}{2} \cos \frac{C}{2} \right\} \right]$$

$$= R^2 \left[1 - 8 \sin \frac{B}{2} \sin \frac{C}{2} \cos \frac{B+C}{2} \right]$$

$$= R^2 \left[1 - 8 \sin \frac{A}{2} \sin \frac{B}{2} \sin \frac{C}{2} \right]$$

$$= R^2 - 2Rr, \text{ on using Theorem 25.1 (ii).}$$

(iii) $OH^2 = OA^2 + AH^2 - 2OA \cdot AH \cos O\hat{A}H$

$$= R^2 + (2R \cos A)^2 - 2R(2R \cos A) \cos(C - B),$$

　　　　from Theorem 26.1 (i) and Theorem 27.1 (i) above,

$$= R^2[1 - 4 \cos A \{ \cos(C - B) + \cos(C + B) \}]$$

$$= R^2[1 - 8 \cos A \cos B \cos C].$$

(iv) $IH^2 = AI^2 + AH^2 - 2AI \cdot AH \cos I\hat{A}H$

$$= \left(4R \sin \frac{B}{2} \sin \frac{C}{2} \right)^2 + (2R \cos A)^2 - 2 \left(4R \sin \frac{B}{2} \sin \frac{C}{2} \right) (2R \cos A)$$

$$\times \cos \frac{C - B}{2}, \text{ from Theorem 25.1 (iii) and Theorem 26.1 (i),}$$

$$= 4R^2 \left[4 \sin^2 \frac{B}{2} \sin^2 \frac{C}{2} + \cos^2 A - 4 \cos A \sin \frac{B}{2} \sin \frac{C}{2} \cos \frac{C - B}{2} \right]$$

$$= 4R^2 \left[4 \sin^2 \frac{B}{2} \sin^2 \frac{C}{2} + \cos^2 A - 4 \cos A \sin \frac{B}{2} \sin \frac{C}{2} \right.$$

$$\left. \times \left(\cos \frac{C}{2} \cos \frac{B}{2} + \sin \frac{C}{2} \sin \frac{B}{2} \right) \right]$$

$$= 4R^2 \left[8 \sin^2 \frac{A}{2} \sin^2 \frac{B}{2} \sin^2 \frac{C}{2} + \cos A (\cos A - \sin B \sin C) \right]$$

$$= 4R^2 \left[8 \sin^2 \frac{A}{2} \sin^2 \frac{B}{2} \sin^2 \frac{C}{2} - \cos A (\cos(B + C) + \sin B \sin C) \right]$$

$$= 4R^2 \left[8 \sin^2 \frac{A}{2} \sin^2 \frac{B}{2} \sin^2 \frac{C}{2} - \cos A \cos B \cos C \right]$$

$$= 2r^2 - 4R^2 \cos A \cos B \cos C, \text{ from Theorem 25.1 (ii).}$$

$\square$

27.2   Problems / exercises

1) Show that $OH^2 = 9R^2 - (a^2 + b^2 + c^2)$.

Solution. $a^2 + b^2 + c^2 = 4R^2(\sin^2 A + \sin^2 B + \sin^2 C)$, by the sine rule.

$$
\begin{aligned}
\text{Now, } \sin^2 A + \sin^2 B + \sin^2 C &= 1 - \cos^2 A + \frac{1 - \cos(2B)}{2} + \frac{1 - \cos(2C)}{2} \\
&= 2 - \cos^2 A - \frac{\cos(2B) + \cos(2C)}{2} \\
&= 2 - \cos^2 A - \cos(B + C)\cos(B - C) \\
&= 2 + \cos A\{\cos(B - C) + \cos(B + C)\}, \\
&\qquad \text{since } \cos(B + C) = -\cos A, \\
&= 2 + 2\cos A \cos B \cos C,
\end{aligned}
$$

$$
\begin{aligned}
\therefore 9R^2 - (a^2 + b^2 + c^2) &= R^2[9 - 8(1 + \cos A \cos B \cos C)] \\
&= R^2[1 - 8\cos A \cos B \cos C] \\
&= OH^2, \text{ from Theorem 27.1 (iii).}
\end{aligned}
$$

Exercise. Prove

(i) $OI^2 = R^2 - \dfrac{abc}{2s}$;

(ii) $\dfrac{r^2}{\Delta} = \tan\dfrac{A}{2}\tan\dfrac{B}{2}\tan\dfrac{C}{2}.$

◇

Chapter 28

External Angle Bisectors and the Ex-centres I_a, I_b, I_c

28.1 The 3 ex-circles; centres I_a, I_b, I_c

Refer to Fig. 28.1.1.

The ex-circles of $\triangle ABC$ have their centres at I_a, I_b, I_c. I_a is the intersection of the bisectors of the exterior angles at B, C. r_a is the radius of the ex-circle drawn outside the $\triangle ABC$. X', Y', Z' are the contact points with the sides BC, AC extended, AB extended.

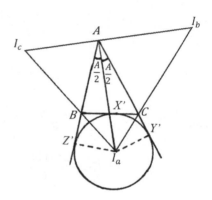

Fig. 28.1.1

28.2 Theorem

Theorem 28.1.

(i) $AY' = AZ' = s$; and $CX' = CY' = s - b, BX' = BZ' = s - c.$

(ii) $II_a = 4R \sin \dfrac{A}{2}$ and $I_b I_c = 4R \cos \dfrac{A}{2}.$

(iii) $r_a = s \tan \dfrac{A}{2} = \dfrac{\Delta}{s - a} = 4R \sin \dfrac{A}{2} \cos \dfrac{B}{2} \cos \dfrac{C}{2}.$

(iv) $\triangle I_a I_b I_c$ *has circumference $2R$, and*

$$\text{area } \triangle I_a I_b I_c = 8R^2 \cos \dfrac{A}{2} \cos \dfrac{B}{2} \cos \dfrac{C}{2}.$$

103

Proof. Note that $I_b A I_c$ is a line perpendicular to $A I I_a$ and that $\triangle ABC$ is the pedal triangle of $\triangle I_a I_b I_c$ so that by Theorem 26.1 (iv) the circumradius of the latter is $2R$. This result is reproved in (iv) below.

(i) Let $BX' = BZ' = x$; let $CX' = CY' = y$.

Then $2AZ' = AZ' + AY' = (c + x) + (b + y) = a + b + c = 2s$.

$\therefore AZ' = AY' = s$; hence $BX' = BZ' = AZ' - c = s - c$ and $CX' = CY' = AY' - b = s - b$.

(ii) Consider $\triangle IBI_a$ where $I\hat{B}I_a = 90°$ and $B\hat{I}I_a = \frac{A}{2} + \frac{B}{2} = 90° - \frac{C}{2}$.

$$II_a = \frac{BI}{\cos B\hat{I}I_a} = \frac{4R \sin \frac{A}{2} \sin \frac{C}{2}}{\sin \frac{C}{2}}, \text{ from Theorem 25.1 (iii),}$$

$$= 4R \sin \frac{A}{2}.$$

$B\hat{I_a}C = \frac{1}{2}Y'\hat{I_a}Z' = \frac{1}{2}(180° - A) = 90° - \frac{A}{2}$, and

$I_a\hat{B}C = \frac{1}{2}Z'\hat{B}X' = 90° - \frac{A}{2}$.

Likewise, $I_a\hat{C}B = \frac{1}{2}Y'\hat{C}X' = 90° - \frac{C}{2}$. It follows that $\triangle I_aCB$ is similar to $\triangle I_a I_b I_c$.

Thus $\dfrac{I_b I_c}{BC} = \dfrac{I_c I_a}{I_a C} = \dfrac{1}{\cos I_c\hat{I_a}C}$,

$$\therefore I_b I_c = \frac{a}{\cos(90° - \frac{A}{2})} = \frac{2R \sin A}{\sin \frac{A}{2}} = 4R \cos \frac{A}{2}.$$

(iii) $r_a = AZ' \tan \dfrac{A}{2}$

$$= s \tan \frac{A}{2}, \text{ from Theorem 28.1 (i)}$$

$$= s\sqrt{\frac{(s-b)(s-c)}{s(s-a)}} \text{ from Theorem 15.1 (iii)}$$

$$= \frac{\Delta}{s-a}, \text{ from Hero's formula.}$$

[Alternatively, from the fact that $\triangle ABC = \triangle ABI_a + \triangle ACI_a - \triangle BCI_a$ on taking their areas, $\Delta = \frac{1}{2}cr_a + \frac{1}{2}br_a - \frac{1}{2}ar_a = \frac{1}{2}r_a(c + b - a) = r_a(s - a)$.]

Also

$$4R \sin \frac{A}{2} \cos \frac{B}{2} \cos \frac{C}{2} = 4R \sqrt{\frac{(s-b)(s-c)}{bc}} \sqrt{\frac{s(s-b)}{ca}} \sqrt{\frac{s(s-c)}{ab}}$$

$$= \frac{4R}{abc} s(s-b)(s-c)$$

$$= \frac{4R}{abc} \frac{\Delta^2}{s-a}$$

$$= \frac{\Delta}{s-a}, \text{ since } \frac{abc}{4R} = \Delta \text{ by Theorem 7.1,}$$

$$= r_a, \text{ by the result just proved.}$$

(iv) For $\triangle I_a I_b I_c$, the circumradius is

$$\frac{1}{2} \frac{I_b I_c}{\sin I_c \widehat{I_a} I_b} = \frac{1}{2} \frac{4R \cos \frac{A}{2}}{\sin(90° - \frac{A}{2})}, \text{ by Theorem 28.1 (ii),}$$

$$= 2R.$$

$$\text{area } \triangle I_a I_b I_c = \frac{1}{2} I_a I_c \cdot I_a I_b \sin I_c \widehat{I_a} I_b$$

$$= \frac{1}{2} \left(4R \cos \frac{B}{2} \right) \left(4R \cos \frac{C}{2} \right) \sin \left(90° - \frac{A}{2} \right)$$

$$= 8R^2 \cos \frac{A}{2} \cos \frac{B}{2} \cos \frac{C}{2}.$$

$\square$

28.3 Problems / exercises

1) Prove the relations,

(i) $\dfrac{1}{r_a} + \dfrac{1}{r_b} + \dfrac{1}{r_c} = \dfrac{1}{r}$;

(ii) $r_a r_b r_c = r s^2$;

(iii) $r_a + r_b + r_c - r = 4R$.

Solution. (i) $LHS = \dfrac{s-a}{\Delta} + \dfrac{s-b}{\Delta} + \dfrac{s-c}{\Delta} = \dfrac{3s - (a+b+c)}{\Delta} = \dfrac{s}{\Delta} = \dfrac{1}{r} = RHS$;

(ii) $LHS = \dfrac{\Delta}{s-a}\dfrac{\Delta}{s-b}\dfrac{\Delta}{s-c} = \dfrac{\Delta^3}{\Delta^2/s} = s\Delta = rs^2 = RHS;$

(iii) Here it is more preferable to use the trigonometrical expressions, by Theorem 28.1 (iii) and Theorem 25.1 (ii),

$$LHS = 4R\{S_A C_B C_C + S_B C_C C_A + S_C C_A C_B - S_A S_B S_C\},$$

where $S_A \equiv \sin\frac{A}{2}$, $C_A \equiv \cos\frac{A}{2}$, etc., it follows that

$$LHS = 4R\{(S_A C_B + S_B C_A)C_C + S_C(C_A C_B - S_A S_B)\}$$
$$= 4R\left\{ \sin\frac{A+B}{2}\cos\frac{C}{2} + \sin\frac{C}{2}\cos\frac{A+B}{2} \right\}$$
$$= 4R\sin\frac{A+B+C}{2}$$
$$= 4R$$
$$= RHS.$$

Exercise. Prove,

(i) $(r_a - r)(r_b - r)(r_c - r) = 4r^2 R;$

(ii) $II_a^2 + I_b I_c^2 = 16R^2;$

(iii) $\Delta = \sqrt{rr_a r_b r_c}.$ ◇

Chapter 29

The Distances AI_a, OI_a, HI_a

29.1   Theorem

Theorem 29.1. *In* $\triangle ABC$,

(i) $AI_a = 4R \cos \dfrac{B}{2} \cos \dfrac{C}{2}$;

(ii) $OI_a^2 = R^2 \left(1 + 8 \sin \dfrac{A}{2} \cos \dfrac{B}{2} \cos \dfrac{C}{2} \right)$

$\qquad = R^2 + 2Rr_a$;

(iii) $HI_a^2 = 2r_a^2 - 4R^2 \cos A \cos B \cos C$.

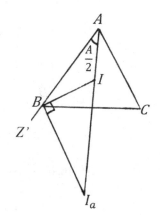

Fig. 29.1.1

Proof. Refer to Fig. 29.1.1.

(i) Consider $\triangle ABI_a$, $B\hat{A}I_a = \frac{A}{2}$,

$A\widehat{I_a}B = Z'\hat{B}I_a - \frac{A}{2} = (90° - \frac{B}{2}) - \frac{A}{2} = \frac{C}{2}$.

By the sine rule (Theorem 6.1),

$$\frac{AI_a}{\sin A\hat{B}I_a} = \frac{c}{\sin A\widehat{I_a}B},$$

i.e.,

$$\frac{AI_a}{\sin(90° + \frac{B}{2})} = \frac{c}{\sin \frac{C}{2}}.$$

$\therefore AI_a = 2R \sin C \dfrac{\cos \frac{B}{2}}{\sin \frac{C}{2}} = 4R \cos \dfrac{B}{2} \cos \dfrac{C}{2}$.

(ii) $OI_a^2 = OA^2 + AI_a^2 - 2OA \cdot AI_a \cos O\hat{A}I_a$

$$= R^2 + \left(4R\cos\frac{B}{2}\cos\frac{C}{2}\right)^2 - 2R\left(4R\cos\frac{B}{2}\cos\frac{C}{2}\right)\cos\frac{C-B}{2},$$

from Theorem 29.1 (i),

$$= R^2\left[1 + 16\cos^2\frac{B}{2}\cos^2\frac{C}{2} - 8\cos\frac{B}{2}\cos\frac{C}{2}\left(\cos\frac{B}{2}\cos\frac{C}{2}\right.\right.$$
$$\left.\left. + \sin\frac{B}{2}\sin\frac{C}{2}\right)\right]$$

$$= R^2\left[1 + 8\cos\frac{B}{2}\cos\frac{C}{2}\left(\cos\frac{B}{2}\cos\frac{C}{2} - \sin\frac{B}{2}\sin\frac{C}{2}\right)\right]$$

$$= R^2\left[1 + 8\cos\frac{B}{2}\cos\frac{C}{2}\cos\frac{B+C}{2}\right]$$

$$= R^2\left[1 + 8\sin\frac{A}{2}\cos\frac{B}{2}\cos\frac{C}{2}\right]$$

$$= R^2 + 2Rr_a, \quad \text{on using Theorem 28.1 (iii).}$$

(iii) $HI_a^2 = AI_a^2 + HA^2 + 2AI_a \cdot HA\cos\frac{B-C}{2}$

$$= \left(4R\cos\frac{B}{2}\cos\frac{C}{2}\right)^2 + (2R\cos A)^2 - 2\left(4R\cos\frac{B}{2}\cos\frac{C}{2}\right)$$
$$\times (2R\cos A)\cos\frac{C-B}{2}$$

$$= R^2\left[16\cos^2\frac{B}{2}\cos^2\frac{C}{2} + 4\cos^2 A - 16\cos A\cos\frac{B}{2}\cos\frac{C}{2}\right.$$
$$\left. \times \left(\cos\frac{B}{2}\cos\frac{C}{2} + \sin\frac{B}{2}\sin\frac{C}{2}\right)\right]$$

$$= R^2\left[16\cos^2\frac{B}{2}\cos^2\frac{C}{2}(1 - \cos A) + 4\cos A(\cos A - \sin B\sin C)\right]$$

$$= R^2\left[32\sin^2\frac{A}{2}\cos^2\frac{B}{2}\cos^2\frac{C}{2} - 4\cos A\cos B\cos C\right]$$

$$= 2\left(4R\sin\frac{A}{2}\cos\frac{B}{2}\cos\frac{C}{2}\right)^2 - 4R^2\cos A\cos B\cos C$$

$$= 2r_a{}^2 - 4R^2\cos A\cos B\cos C.$$

$\square$

29.2 Problems / exercises

1) Prove that $4rr_a = II_a^2 \sin B \sin C$.

Solution. $4rr_a = 4\dfrac{\Delta}{s}\dfrac{\Delta}{s-a} = \dfrac{4\Delta^2}{s(s-a)} = 4(s-b)(s-c);$

$$II_a = AI_a - II_a$$

$$= 4R\cos\frac{B}{2}\cos\frac{C}{2} - 4R\sin\frac{B}{2}\sin\frac{C}{2},$$

from Theorem 29.1 (i) and Theorem 28.1 (ii),

$$= 4R\cos\frac{B+C}{2}$$

$$= 4R\sin\frac{A}{2}.$$

$\therefore II_a^2 = 16R^2\sin^2\dfrac{A}{2} = 16R^2\dfrac{(s-b)(s-c)}{bc}$ by Theorem 15.1 (i).

Hence, we have

$$II_a^2 \sin B \sin C = 16R^2(s-b)(s-c)\frac{\sin B}{b}\frac{\sin C}{c}$$

$$= 4(s-b)(s-c)$$

$$= 4rr_a.$$

The Nine-Point Centre, N. The Feuerbach Circle

30.1 Introduction

In $\triangle ABC$, let A', B', C' be the midpoints of the sides; let D, E, F be the feet of the altitudes from the vertices; let H be the orthocentre. Denote by P, Q, R the midpoints of AH, BH, CH. We prove below that all 9 points $(A', B', C', D, E, F, P, Q, R)$ lie on a circle — the Feuerbach circle. This circle circumscribes $\triangle DEF$, the pedal triangle and, as such, has a radius $\frac{1}{2}R$ [see Theorem 26.1 (v)]. The centre of the circle is N, the nine-point centre.

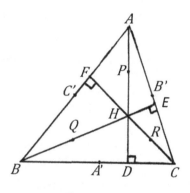

Fig. 30.1.1

30.2 Feuerbach's theorem [H. W. Feuerbach, 1800–1834]

Theorem 30.1 (Feuerbach's theorem). *A circle circumscribes all 9 points $A', B', C', D, E, F, P, Q, R$.*

Proof. Refer to Fig. 30.1.1.

All 9 points are shown $PC' \parallel BE$, so $PC' \perp AC$.

$A'C' \parallel AC$, so $\widehat{PC'A'} = 90°$.

Similarly, $\widehat{PB'A'} = 90°$, so PA' is a diameter of the circle through A', B', C'.

In the same way QB', RC' are diameters of the same circle through the midpoints A', B', C'. Since $\widehat{PDA'} = 90°$, $\widehat{QEB'} = 90°$, $\widehat{RFC'} = 90°$, the circle passes through D, E, F as well. $\qquad \square$

The centre of the Feuerbach circle lies at the intersection of the perpendicular bisectors of the chords $A'D$, $B'E$, $C'F$.

30.3   The Euler line [Leonhard Euler, 1707–1783]

Theorem 30.2 (The Euler line). *The circumcentre O, the centroid G, the nine-point centre N and the orthocentre H of $\triangle ABC$ are collinear. Further, $OG : GN : NH = 2 : 1 : 3$.*

Proof. Refer to Fig. 30.3.1.

OA', AH are each $\perp BC$,

$\therefore OA' \parallel AH$.

G lies on AA' with $\frac{AG}{GA'} = 2$.

But, $\frac{AH}{OA'} = \frac{2R\cos A}{R\cos A} = 2$.

$\therefore OGH$ is a straight line and $\frac{OG}{GH} = \frac{1}{2}$.

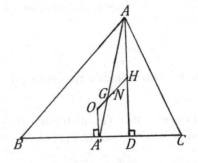

Fig. 30.3.1

Now, N lies on the perpendicular bisectors of all the three chords $A'D, B'E, C'F$ of the nine-point circle. These all bisect ON, i.e. $ON = NH$, so finally $OG : GN : NH = 2 : 1 : 3$. $\qquad \square$

30.4 Problems / exercises

1) From the circumcentre O, the vectors $\overline{OA} = \underline{a}, \overline{OB} = \underline{b}, \overline{OC} = \underline{c}$ to the vertices are drawn.

Show that $\overline{OG} = \frac{1}{3}(\underline{a} + \underline{b} + \underline{c})$ and find $\overline{ON}, \overline{OH}$ in terms of $\underline{a}, \underline{b}, \underline{c}$.

Solution. A' being the mid-point of BC, $\overline{OA'} = \frac{1}{2}(\underline{b} + \underline{c})$,

$\overline{AA'} = \frac{1}{2}(\underline{b} + \underline{c}) - \underline{a}$,

so $\overline{OG} = \underline{a} + \frac{2}{3}\overline{AA'} = \underline{a} + \frac{2}{3}\left(\frac{1}{2}(\underline{b} + \underline{c}) - \underline{a}\right) = \frac{1}{3}(\underline{a} + \underline{b} + \underline{c})$.

Next, $\overline{ON} = \frac{3}{2}\overline{OG}$, from Theorem 30.2.

So, $\overline{ON} = \frac{1}{2}(\underline{a} + \underline{b} + \underline{c})$, and $\overline{OH} = 3\overline{OG} = \underline{a} + \underline{b} + \underline{c}$.

Chapter 31

The Distances IG, IN

31.1 Theorem

Theorem 31.1. O, G, N, H *being on the Euler line,*

(i) $IG^2 = \dfrac{2}{9}\{3(R-r)^2 - R^2(1 - 2\cos A \cos B \cos C)\}$;

(ii) $IN = \dfrac{1}{2}R - r$, *and hence the nine-point circle of* $\triangle ABC$ *touches the incircle (Feuerbach).*

Proof. Refer to Fig. 31.1.1.

(i) We use the facts that

$OH^2 = R^2(1 - 8\cos A \cos B \cos C)$ and

$OG = \frac{1}{3}OH$, from Theorem 27.1 (iii) and Theorem 30.2.

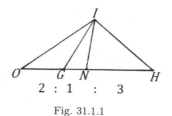

Fig. 31.1.1

Also,
$IH^2 = 2r^2 - 4R^2 \cos A \cos B \cos C,$
$IO^2 = R^2 - 2rR,$ from Theorem 27.1 (iv) and Theorem 27.1 (ii).

$$IH^2 = IG^2 + HG^2 - 2\,IG \cdot HG \cos I\hat{G}H$$
$$= IG^2 + (2\,OG)^2 + 2\,IG \cdot (2\,OG)\cos I\hat{G}O \ldots\ldots (1)$$
$$IO^2 = IG^2 + OG^2 - 2\,IG \cdot OG \cos I\hat{G}O \ldots\ldots\ldots\ldots (2)$$

Eliminating the trigonometric term from (1) and (2):
$$2\,IO^2 + IH^2 = 3\,IG^2 + 6\,OG^2.$$

Hence, $3\,IG^2 = IH^2 + 2\,IO^2 - 6\,OG^2$

$$= IH^2 + 2\,IO^2 - \frac{2}{3}\,OH^2$$

$$= (2r^2 - 4R^2\cos A\cos B\cos C) + 2(R^2 - 2rR)$$

$$- \frac{2}{3}R^2(1 - 8\cos A\cos B\cos C)$$

$$= 2(R - r)^2 - \frac{2}{3}R^2(1 - 2\cos A\cos B\cos C),$$

and thus (i) is proved.

(ii) Refer to Fig. 31.1.2.

By Apollonius' theorem

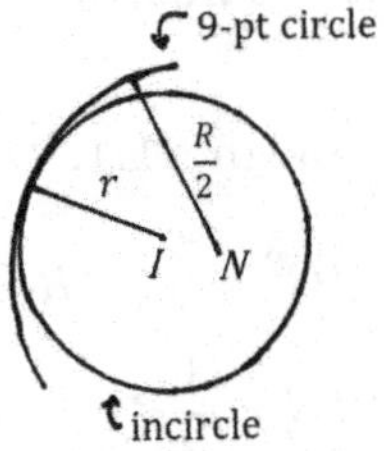

Fig. 31.1.2

$$IO^2 + IH^2 = 2IN^2 + 2HN^2$$

$$= 2IN^2 + \frac{1}{2}OH^2, \quad (\because NH = \frac{1}{2}OH)$$

$$\therefore 2IN^2 = (R^2 - 2rR) + (2r^2 - 4R^2\cos A\cos B\cos C)$$

$$- \frac{1}{2}R^2(1 - 8\cos A\cos B\cos C)$$

$$= \frac{1}{2}R^2 - 2rR + 2r^2$$

$$= \frac{1}{2}(R - 2r)^2.$$

$\therefore IN^2 = \left(\frac{R}{2} - r\right)^2$, $IN = \left|\frac{R}{2} - r\right|$, and hence the Feuerbach result: that the nine-point circle touches the incircle.

$\square$

Chapter 32

Napoleon Circles and the Fermat Point

32.1 Introduction

Suppose $\triangle ABC$ is such that none of its angles exceeds $120°$. Draw equilateral triangles BCX, CAY, ABZ exterior to it, with A_1, B_1, C_1 as their circumcentres. The triangle $\triangle A_1 B_1 C_1$ is called an (outer) Napoleon triangle. The three triangles are Napoleon triangles. (Napoleon Bonaparte, besides being a brilliant general, was also a keen geometer). We prove below, that $\triangle A_1 B_1 C_1$ is equilateral; that AX, BY, CZ are concurrent and are of equal length; and, further, that the point of concurrency, F — the Fermat point, has another special property.

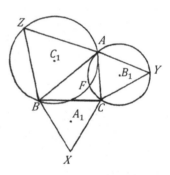

Fig. 32.1.1

32.2 Napoleon's theorem [Napoleon Bonaparte, 1769–1821]

Theorem 32.1 (Napoleon's Theorem). Refer to Fig. 32.1.1.

(i) $AX = BY = CZ$; the lines AX, BY, CZ are inclined to each other at $60°$ and intersect at F, the Fermat point.

(ii) $\triangle A_1 B_1 C_1$ is equilateral.

Proof. (i) Compare $\triangle ZBC$ with $\triangle ABX$: they have a common vertex B and are congruent, as the former is obtained from the latter by a $60°$ anti-clockwise rotation about B. Hence $CZ = AX$.

Similarly we find $AX = YB$.

Also, on account of the $60°$ rotations, the three lines cut one another at $60°$. Consider the circles ABZ, ACY. They meet at a point F.

$A\hat{F}B = 180° - Z = 120°$ and $A\hat{F}C = 180° - Y = 120°$.

$\therefore B\hat{F}C = 120°$ also; so $BFCX$ is cyclic and so $B\hat{F}X = B\hat{C}X = 60°$.

Hence $A\hat{F}X = A\hat{F}B + B\hat{F}X = 180°$.

Thus AX is a straight line through F. Equally then, BFY, CFZ are straight lines. F is the Fermat point of $\triangle ABC$.

We work out an expression for BY (for instance): From $\triangle ABC$

$$BY^2 = c^2 + b^2 - 2bc\cos(A + 60°)$$
$$= c^2 + b^2 - 2bc\left(\frac{1}{2}\cos A - \frac{\sqrt{3}}{2}\sin A\right)$$
$$= c^2 + b^2 - \frac{1}{2}(b^2 + c^2 - a^2) + 2\sqrt{3}\Delta$$
$$= \frac{1}{2}(a^2 + b^2 + c^2) + 2\sqrt{3}\Delta,$$

a symmetric expression, which shows that $BY = CZ = AX$.

(ii) $AC_1 = \frac{c}{\sqrt{3}}, AB_1 = \frac{b}{\sqrt{3}}$.

So, from $\triangle AB_1C_1$,

$$B_1C_1^2 = AC_1^2 + AB_1^2 - 2AC_1 \cdot AB_1 \cos C_1\hat{A}B_1$$
$$= \frac{1}{3}c^2 + \frac{1}{3}b^2 - \frac{2}{3}bc\cos(A + 60°)$$
$$= \frac{1}{3}BY^2,$$

from the calculation in (i).

Hence, $B_1C_1 = \frac{1}{\sqrt{3}}BY (= \frac{1}{\sqrt{3}}AX = \frac{1}{\sqrt{3}}CZ)$, and $\triangle A_1B_1C_1$ is equilateral. $\qquad\square$

32.3 The Fermat point, F [Pierre de Fermat, 1601–1665]

Theorem 32.2 (The Fermat Point). *The point F is the unique point inside $\triangle ABC$ to make $FA + FB + FC$ minimal.*

Proof. Refer to Fig. 32.3.1.

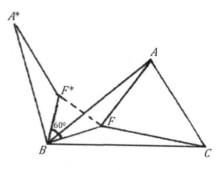

Initially let F be any point in the triangle. Rotate $\triangle AFB$ about B through 60° anti-clockwise, into the position A^*F^*B. $\triangle BFF^*$ is equilateral,

so $FA + FB + FC = F^*A^* + F^*F + FC$.

The point A^* is fixed relative to $\triangle ABC$ and so to C in particu-lar. The least value of the above

Fig. 32.3.1

sum will occur when A^*, F^*, F, C are aligned, i.e. when F, F^* lie on A^*C.

In that case, $B\hat{F}C = 180° - F^*\hat{F}B = 120°$,

and $A\hat{F}B = A\widehat{F^*}B = 180° - B\widehat{F^*}F = 120°$.

Thus, $A\hat{F}C = 60°$.

(The point F is easily constructed: we draw the Napoleon equilateral tri-angles outside $\triangle ABC$. Their circumcircles will intersect at F.)

$\square$

Chapter 33

Small Angles in Radians

Remark Here and in all succeeding chapters, all angles will be in RADI-ANS. (Look back to Chapter 2.)

33.1 Theorem

Theorem 33.1. *If* $0 < \theta < \frac{\pi}{2}$, *then*

$$\sin \theta < \theta < \tan \theta.$$

Proof.
Refer to Fig. 33.1.1.

OAR is a sector of a circle. AT is tangent at A meeting OR at T.

Let r =radius of arc AR; $\theta = A\hat{O}R$

Consider areas:

area $\triangle OAR <$ area sector $OAR <$ area $\triangle OAT$,

$\therefore \dfrac{1}{2}r^2 \sin \theta < \dfrac{1}{2}r^2 \theta < \dfrac{1}{2}r^2 \tan \theta$, (see Chapter 2.)

$\therefore \sin \theta < \theta < \tan \theta$.

This is a fundamental inequality for acute angles. $\square$

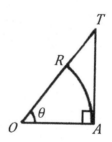

Fig. 33.1.1

121

33.2   Theorem

Theorem 33.2. *If* $0 < \theta < \frac{\pi}{2}$ *and* θ *decreases (through positive values) indefinitely to 0* $(\theta \to 0+)$, *then* $\frac{\theta}{\sin\theta}$ *decreases towards 1;* $\frac{\theta}{\tan\theta}$ *increases towards 1. In short,*

$$\lim_{\theta \to 0+} \frac{\theta}{\sin\theta} = 1; \lim_{\theta \to 0+} \frac{\theta}{\tan\theta} = 1.$$

Proof. From Theorem 33.1,

$$1 < \frac{\theta}{\sin\theta} < \frac{1}{\cos\theta} \; (\theta \neq 0).$$

As $\theta \to 0+$, $\cos\theta$ increases towards 1, hence $\frac{\theta}{\sin\theta}$ must decrease towards 1.

Again, $\cos\theta < \dfrac{\theta}{\tan\theta} < 1.$ $(\theta \neq 0)$

So $\frac{\theta}{\tan\theta}$ must increase towards 1, with the help of some calculus.   □

33.3   Problems / exercises

1) For $\theta = 0.05, 0.01, 0.002$, evaluate $\frac{\theta}{\sin\theta}, \frac{\theta}{\tan\theta}$, verifying Theorem 33.2.

Solution. We exhibit the results in a table:

	$\theta = 0.05$	$\theta = 0.01$	$\theta = 0.002$
$\frac{\theta}{\sin\theta} =$	1.000416708	1.000016667	1.000000667
$\frac{\theta}{\tan\theta} =$	0.9991665277	0.9999666664	0.99999866666

The tendencies are now clear.

Chapter 34

Euler's Product. A Limit for π

34.1 Euler's limit [Leonhard Euler, 1707–1783]

Theorem 34.1.

$$\lim_{n \to \infty} \left(\cos\frac{\theta}{2} \cos\frac{\theta}{4} \cdots \cos\frac{\theta}{2^n} \right) = \frac{\sin\theta}{\theta}, \qquad (34.1)$$

or briefly, $\displaystyle\lim_{n \to \infty} \prod_{r=1}^{n} \cos\frac{\theta}{2^r} = \frac{\sin\theta}{\theta}$.

Proof. We start with the identities:

$$\cos\frac{\theta}{2} = \frac{\sin\theta}{2\sin\frac{\theta}{2}}, \quad \cos\frac{\theta}{4} = \frac{\sin\frac{\theta}{2}}{2\sin\frac{\theta}{4}}, \quad \cos\frac{\theta}{8} = \frac{\sin\frac{\theta}{4}}{2\sin\frac{\theta}{8}}, \quad \cdots$$

Multiplying n of these together gives

$$\cos\frac{\theta}{2} \cos\frac{\theta}{4} \cdots \cos\frac{\theta}{2^n} = \frac{\sin\theta}{2^n \sin\frac{\theta}{2^n}},$$

after cancelling down all but two of the sine terms on the RHS. (The product 'telescopes' down).

Now

$$2^n \sin\frac{\theta}{2^n} \equiv \theta\frac{\sin\frac{\theta}{2^n}}{\frac{\theta}{2^n}},$$

and it has limit θ as $n \to \infty$ and $\frac{\theta}{2^n}$ approaches 0, by Theorem 33.2.

$$\therefore \lim_{n \to \infty} \left(\cos\frac{\theta}{2} \cos\frac{\theta}{4} \cdots \cos\frac{\theta}{2^n} \right) = \frac{\sin\theta}{\theta}. \qquad \square$$

Corollary 34.1. *Substitute* $\theta = \frac{\pi}{2}$ *into (34.1) as a special case:*

$$\cos\frac{\pi}{4} = \frac{\sqrt{2}}{2},$$

and from $\cos\frac{\theta}{2} = \sqrt{\frac{1+\cos\theta}{2}}$, *we sequentially obtain*

$$\cos\frac{\pi}{8} = \sqrt{\frac{1 + \frac{\sqrt{2}}{2}}{2}} = \frac{\sqrt{2+\sqrt{2}}}{2},$$

$$\cos\frac{\pi}{16} = \sqrt{\frac{1 + \frac{\sqrt{2+\sqrt{2}}}{2}}{2}} = \frac{\sqrt{2+\sqrt{2+\sqrt{2}}}}{2},$$

$$\cos\frac{\pi}{32} = \frac{\sqrt{2+\sqrt{2+\sqrt{2+\sqrt{2}}}}}{2},$$

$$\vdots$$

Hence, Theorem 34.1 gives

$$\frac{\sqrt{2}}{2}\frac{\sqrt{2+\sqrt{2}}}{2}\frac{\sqrt{2+\sqrt{2+\sqrt{2}}}}{2}\cdots = \frac{2}{\pi} = 0.6366197724...$$

34.2 Problems / exercises

1) Evaluate $\cos\frac{\pi}{4}\cos\frac{\pi}{8}\cos\frac{\pi}{16}\cdots\cos\frac{\pi}{1024}$.

Solution. The 9-term product is $0.6366207711...$

Exercise. Show that the 12-term product
$$\cos\frac{\pi}{4}\cos\frac{\pi}{8}\cdots\cos\frac{\pi}{4096}\cos\frac{\pi}{8192} = 0.636619788...$$

◇

Chapter 35

Angle of Dip and Distance to Horizon

35.1 Dip angle to the horizon

Refer to Fig. 35.1.1.

A is a position on the Earth's surface. AP is the vertical through A; $AP = h$; this could possibly represent the height of a mountain. PQ is the distance to the horizon.

The Earth's radius $R \doteqdot 6378$ km. The dip angle $H\hat{P}Q = A\hat{O}Q = \theta$.

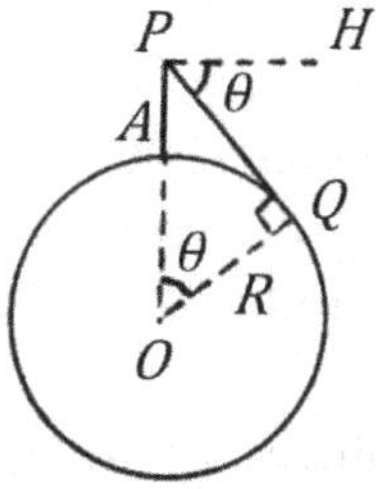

We have

Fig. 35.1.1

$$\cos\theta = \frac{R}{R+h}, \qquad (35.1)$$

$$\begin{aligned}
PQ^2 &= OP^2 - OQ^2 \\
&= (R+h)^2 - R^2 \\
&= 2Rh + h^2 \\
&\doteqdot 2Rh, \text{ if } h \text{ is small compared with } R.
\end{aligned}$$

So

$$PQ \doteqdot \sqrt{2Rh}, \qquad (35.2)$$

and

$$\tan\theta = \frac{PQ}{R} \doteqdot \sqrt{\frac{2h}{R}}. \qquad (35.3)$$

35.2   Problems / exercises

1) Show that the distance to the horizon from a point on a cliff-top h metres high above ground $\simeq 3.57\sqrt{h}\,$km and that the dip to the horizon $\simeq \frac{\sqrt{h}}{31.17}$ degrees. (Take $R = 6378\,$km)

Solution. From (35.2) above, distance to horizon $\simeq \sqrt{2 \times \dfrac{h}{1000} \times 6378} \simeq$ $3.57\sqrt{h}\,$km.

From (35.3), the dip is then found from $\tan\theta \simeq \sqrt{\dfrac{2h}{1000 \times 6378}} = \dfrac{\sqrt{h}}{1786}$.

Because θ is small, $\tan\theta \simeq \theta$ (see Theorem 33.2), so

$$\theta \simeq \frac{\sqrt{h}}{1786} \text{ rads} \simeq \frac{\sqrt{h}}{31.17} \text{ degrees.}$$

2) From a hot-air balloon drifting at a height of 1km above the Earth's surface, an area of nearly $40,000\,$km^2 should be visible. Estimate the Earth's radius from this information.

Solution. Approximately, $\pi(R\sin\theta)^2 = 40,000$; and $\cos\theta = \frac{R}{R+1}$, since $h = 1$ from (35.1).

Eliminating θ, the dip angle, $\pi R^2\left\{1 - \left(\dfrac{R}{R+1}\right)^2\right\} \simeq 40,000$;

$$\frac{\pi R^2(2R+1)}{(R+1)^2} \simeq 40,000.$$

Now, $\dfrac{2R+1}{(R+1)^2} \simeq \dfrac{2R}{R^2} = \dfrac{2}{R}$.

This further approximation gives $2\pi R \simeq 40,000$, $R \simeq 6,366\,$km.

Chapter 36

A Series for π

36.1 Problem

1) Prove (i) $\cot\alpha - 2\cot 2\alpha = \tan\alpha$.

(ii) $\dfrac{1}{2}\tan\dfrac{\alpha}{2} + \dfrac{1}{4}\tan\dfrac{\alpha}{4} + \dfrac{1}{8}\tan\dfrac{\alpha}{8} + \dfrac{1}{16}\tan\dfrac{\alpha}{16} + \cdots = \dfrac{1}{\alpha} - \cot\alpha$.

Consider the special case $\alpha = \frac{\pi}{2}$ in (ii).

Solution. (i) $\begin{aligned}[t]\cot\alpha - 2\cot 2\alpha &= \frac{1}{\tan\alpha} - \frac{2}{\tan 2\alpha} \\ &= \frac{1}{\tan\alpha} - \frac{1 - \tan^2\alpha}{\tan\alpha} \\ &= \tan\alpha.\end{aligned}$

(ii) From (i),

$$\frac{1}{2}\tan\frac{\alpha}{2} = \frac{1}{2}\cot\frac{\alpha}{2} - \cot\alpha. \tag{36.1}$$

Halving this equation and replacing α by $\frac{\alpha}{2}$ gives

$$\frac{1}{4}\tan\frac{\alpha}{4} = \frac{1}{4}\cot\frac{\alpha}{4} - \frac{1}{2}\cot\frac{\alpha}{2}. \tag{36.2}$$

Repeating this process

$$\frac{1}{8}\tan\frac{\alpha}{8} = \frac{1}{8}\cot\frac{\alpha}{8} - \frac{1}{4}\cot\frac{\alpha}{4}. \tag{36.3}$$

and again,

$$\frac{1}{16}\tan\frac{\alpha}{16} = \frac{1}{16}\cot\frac{\alpha}{16} - \frac{1}{8}\cot\frac{\alpha}{8}. \tag{36.4}$$

When n of these equations are added, the sum on the RHS 'telescopes' down. All but two of the terms cancel out, with the result

$$\frac{1}{2}\tan\frac{\alpha}{2} + \frac{1}{4}\tan\frac{\alpha}{4} + \cdots + \frac{1}{2^n}\tan\frac{\alpha}{2^n} = \frac{1}{2^n}\cot\frac{\alpha}{2^n} - \cot\alpha.$$

Now,

$$\frac{1}{2^n}\cot\frac{\alpha}{2^n} = \frac{\frac{1}{2^n}}{\tan(\frac{\alpha}{2^n})} = \frac{1}{\alpha}\frac{\frac{\alpha}{2^n}}{\tan(\frac{\alpha}{2^n})},$$

and as n increases indefinitely this term in the sum tends towards $\frac{1}{\alpha}$, by Theorem 33.2. Thus taking limits in the sum, we obtain (ii).

The special case $\alpha = \frac{\pi}{2}$ produces

$$\frac{2}{\pi} = \frac{1}{2}\tan\frac{\pi}{4} + \frac{1}{4}\tan\frac{\pi}{8} + \frac{1}{8}\tan\frac{\pi}{16} + \cdots , \qquad (36.5)$$

or

$$\pi = \frac{1}{\frac{1}{4}\tan\frac{\pi}{4} + \frac{1}{8}\tan\frac{\pi}{8} + \frac{1}{16}\tan\frac{\pi}{16} + \cdots}.$$

Exercise. Take the first 9 terms of the last series (36.5) and compare the total with $\frac{1}{\pi} = 0.3183098862....$

<u>Ans</u>: Total of the first 9 terms $= 0.31831$ to 5 d.p.   ◇

Chapter 37

Stricter Inequalities. More Limits

37.1 Theorem

Theorem 37.1. *If* $0 < \theta < \frac{\pi}{2}$, *then*

(i) $\theta - \dfrac{\theta^3}{4} < \sin\theta < \theta,$

(ii) $1 - \dfrac{\theta^2}{2} < \cos\theta < 1 - \dfrac{\theta^2}{2} + \dfrac{\theta^4}{16}.$

Proof. (i) The RH inequality, $\sin\theta < \theta$, follows from Theorem 33.2.

Again from Theorem 33.2, $\tan\frac{\theta}{2} > \frac{\theta}{2}$ which gives $\sin\frac{\theta}{2} > \frac{\theta}{2}\cos\frac{\theta}{2}$.

On multiplying the latter by $2\cos\frac{\theta}{2}$:

$$\sin\theta > \theta\cos^2\frac{\theta}{2} = \theta\left(1 - \sin^2\frac{\theta}{2}\right) > \theta\left\{1 - \left(\frac{\theta}{2}\right)^2\right\} = \theta - \frac{\theta^3}{4},$$

which is the LH inequality.

(ii) $\cos\theta = 1 - 2\sin^2\dfrac{\theta}{2} > 1 - 2\left(\dfrac{\theta}{2}\right)^2 = 1 - \dfrac{\theta^2}{2}$. This is the LH inequality.

From (i) $\sin\dfrac{\theta}{2} > \dfrac{\theta}{2} - \dfrac{(\frac{\theta}{2})^3}{4} = \dfrac{\theta}{2} - \dfrac{\theta^3}{32}$, so,

$$\cos\theta < 1 - 2\left(\frac{\theta}{2} - \frac{\theta^3}{32}\right)^2 < 1 - \frac{\theta^2}{2} + \frac{\theta^4}{16},$$

which is the RH inequality. $\qquad\square$

Other methods could be used to provide stricter bounds for $\sin\theta$, $\cos\theta$. We may thereby replace (i), (ii) by

(i)$'$ $\theta - \dfrac{\theta^3}{6} < \sin\theta < \theta$;

(ii)$'$ $1 - \dfrac{\theta^2}{2} < \cos\theta < 1 - \dfrac{\theta^2}{2} + \dfrac{\theta^4}{24}$.

37.2   Huyghens' rule [Christian Huyghens, 1651]

Refer to Fig. 37.2.1.

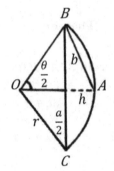

Theorem 37.2 (Huyghens' Rule). *BAC is a circular arc of radius r. Let $B\hat{O}C = \theta$. Put $BC = a$, $BA = b$. Then*

$$\text{arc } BAC \simeq \frac{8b - a}{3}.$$

Fig. 37.2.1

Proof. Arc $BAC = r\theta$, $a = 2r\sin\frac{\theta}{2}$, $b = 2r\sin\frac{\theta}{4}$,

$$\frac{8b - a}{3} = \frac{2r}{3}\left\{ 8\sin\frac{\theta}{4} - \sin\frac{\theta}{2} \right\}$$

$$= \frac{2r}{3}\left\{ 8\left(\frac{\theta}{4} - \frac{(\frac{\theta}{4})^3}{6}\right) - \left(\frac{\theta}{2} - \frac{(\frac{\theta}{2})^3}{6}\right) \right\},$$

(using approximation (i)$'$ in 37.1.)

$$= \frac{2r}{3}\left\{ 2\theta - \frac{\theta^3}{48} - \left(\frac{\theta}{2} - \frac{\theta^3}{48}\right) \right\}$$

$$= r\theta$$

$$= \text{arc } BAC.$$

$\square$

Exercise. If h is the height of the segment BAC in Fig. 37.2.1, show that when θ is very small, the area of this segment $\simeq \frac{2}{3}ha$. $\diamond$

37.3 Problems / exercises

1) Use inequality (ii) in Theorem 37.1 to show that $\cos 0.05 = 0.998750$ correct to 6 d.p.

Solution. Inequality (ii) shows that $\cos 0.05$ lies between $1 - \frac{0.05^2}{2}$ and $1 - \frac{0.05^2}{2} + \frac{0.05^4}{16}$, i.e. between 0.99875 and 0.9987503906.

Thus $\cos 0.05 = 0.998750$ correct to 6 d.p.

Exercise. Use inequality (ii)' to show that $\cos 0.01$ can be given accurately to 10 d.p.

<u>Ans:</u> 0.9999500004. ◇

2) Find

(i) $\displaystyle \lim_{\theta \to 0} \frac{1 - \cos \theta}{\theta^2}$;

(ii) $\displaystyle \lim_{\theta \to \alpha} \frac{\cos \alpha - \cos \theta}{\theta^2 - \alpha^2}$.

Solution. (i) $\displaystyle \frac{1 - \cos \theta}{\theta^2} = \frac{2\sin^2 \frac{\theta}{2}}{\theta^2} = \frac{1}{2}\left(\frac{\sin \frac{\theta}{2}}{\frac{\theta}{2}}\right)^2$.

Its limit, as $\theta \to 0$, is $\frac{1}{2}$.

(ii) $\displaystyle \frac{\cos \alpha - \cos \theta}{\theta^2 - \alpha^2} = \frac{\cos \alpha - \cos(\alpha + h)}{(\alpha + h)^2 - \alpha^2}$,

on putting $\theta = \alpha + h$,

$$= \frac{2\sin(\alpha + \frac{h}{2})\sin \frac{h}{2}}{2\alpha h + h^2}$$

$$= 2\sin\left(\alpha + \frac{h}{2}\right)\frac{\sin \frac{h}{2}}{\frac{h}{2}}\frac{1}{4\alpha + 2h}.$$

Since $\theta \to \alpha$, so $h \to 0$ and the limit of the RH expression is

$$2\sin \alpha \cdot 1 \cdot \frac{1}{4\alpha} = \frac{\sin \alpha}{2\alpha}.$$

Exercise.

(i) A circle of radius r, circumscribes the vertices of a regular n-sided polygon. Show that the area of the polygon is $r^2 n \sin \frac{\pi}{n} \cos \frac{\pi}{n}$.

(ii) A circle, radius r, is the incircle to a regular n-sided polygon. Show that the area of this polygon is $r^2 n \tan \frac{\pi}{n}$.

(iii) What happens as $n \to \infty$ in (i) and (ii)?

Ans: Both areas approach πr^2.   ◇

Appendix A

Elementary Spherical Trigonometry

A.1 Latitude and longitude

Refer to Fig. A.1.1.

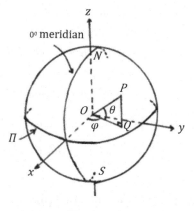

The nearly spherical Earth, mean radius R, is shown, with its centre at O, the N. and S. Poles, the equatorial plane Π and the $0°$-meridian line through Greenwich.

Rectangular axes $Oxyx$ are taken, with Ox cutting the meridian and Π, and Oz through N.

P is taken as a general point on the Earth's surface and OQ is the projection of OP on Π: $O\hat{Q}P = \frac{\pi}{2}$.

Fig. A.1.1

$P\hat{O}Q = \theta$ is the latitude of P; $Q\hat{O}x = \varphi$ is the longitude of P.

$-\frac{\pi}{2} \leqslant \theta \leqslant \frac{\pi}{2}$ and $-\pi < \varphi \leqslant \pi$.

$\theta = 0$ corresponds to the equator, and $\varphi = 0$ corresponds to the Greenwich meridian. P has coordinates (x, y, z) where $x = OQ \cos\varphi = R \cos\theta \cos\varphi$, $y = OQ \sin\varphi = R \cos\theta \sin\varphi$, $z = QP = R \sin\theta$, since $OP = R$.

Note: The Earth's equatorial radius is $6378 \, \text{km} \simeq R$.

A.2   Shortest distances over the surface

Refer to Fig. A.2.1.

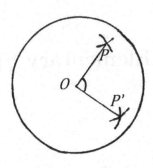

P, P' are taken with coordinates (x, y, z), (x', y', z') respectively, and corresponding latitudes, longitudes of (θ, φ), (θ', φ'). The plane OPP' cuts the surface in an arc PP' whose length is the shortest (geodesic) distance between the positions.

$P\hat{O}P'$ is found from the scalar (dot) product:

Fig. A.2.1

$$\overline{OP} \bullet \overline{OP'} = R^2 \cos P\hat{O}P',$$

$$\begin{aligned}
\overline{OP} \bullet \overline{OP'} &= (x, y, z) \bullet (x', y', z') \\
&= xx' + yy' + zz' \\
&= R^2(\cos\theta\cos\varphi, \cos\theta\sin\varphi, \sin\theta) \bullet (\cos\theta'\cos\varphi', \cos\theta'\sin\varphi', \sin\theta') \\
&= R^2[\cos\theta\cos\theta'(\cos\varphi\cos\varphi' + \sin\varphi\sin\varphi') + \sin\theta\sin\theta'] \\
&= R^2[\cos\theta\cos\theta'(\cos(\varphi - \varphi')) + \sin\theta\sin\theta'].
\end{aligned}$$

Thus $P\hat{O}P' = \cos^{-1}[\cos\theta\cos\theta'(\cos(\varphi - \varphi') + \sin\theta\sin\theta']$, after which,

$$\text{arc}\, PP' = R \cdot P\hat{O}P', \quad \text{where } P\hat{O}P' \text{ is in radians.} \tag{A.1}$$

A.3   Cosine rule for spherical triangles

Refer to Fig. A.3.1.

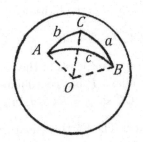

ABC is a spherical triangle, i.e., the arcs AB, BC, CA are geodesics between the respective points on the sphere. The angles of the spherical triangle ABC are denoted by A, B, C and defined to be $A = $ angle between the planes AOB, AOC, etc.

The arcs AB, BC, CA along great circles represent shortest distances.

Fig. A.3.1

If the sphere's radius is 1, the side lengths in the triangle are: $a = B\hat{O}C$, $b = A\hat{O}C$, $c = A\hat{O}B$.

The sides of an actual triangle on the Earth's surface are Ra, Rb, Rc (the angles A, B, C are unaltered).

Cosine rule

Theorem A.3 (Cosine rule). *In any spherical triangle ABC*

$$\cos A = \frac{\cos a - \cos b \cos c}{\sin b \sin c}; \ i.e., \ \cos a = \cos A \sin b \sin c + \cos b \cos c,$$

with a similar result in respect of each of the other two vertices.

Proof. Refer to Fig. A.3.2.

Assume $R = 1$ for our purpose. Slide the actual triangle over the surface, so that A is at the N. pole and AB lies on the $0°$-meridian line.

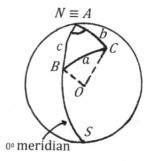

Let θ, θ' be the latitudes of B, C.

Then $c = A\hat{O}B = \frac{\pi}{2} - \theta$, $b = A\hat{O}C = \frac{\pi}{2} - \theta'$ and $a = B\hat{O}C$.

The difference of the longitudes of B, C is the angle A.

Fig. A.3.2

From A.2, $\cos a = \cos B\hat{O}C$

$$= \cos \theta' \cos \theta \cos A + \sin \theta' \sin \theta$$
$$= \sin b \sin c \cos A + \cos b \cos c,$$

or recasting,

$$\cos A = \frac{\cos a - \cos b \cos c}{\sin b \sin c}.$$

□

A.4   Sine rule

Theorem A.4 (Sine rule). *In the spherical triangle* ABC,
$$\frac{\sin A}{\sin a} = \frac{\sin B}{\sin b} = \frac{\sin C}{\sin c}.$$

Proof.

$$\sin^2 A = 1 - \left(\frac{\cos a - \cos b \cos c}{\sin b \sin c}\right)^2, \text{ from the cosine rule in A.3.}$$

$$= \frac{1}{\sin^2 b \sin^2 c}\left\{(1 - \cos^2 b)(1 - \cos^2 c) - (\cos a - \cos b \cos c)^2\right\}$$

$$= \frac{1}{\sin^2 b \sin^2 c}\left\{1 - \cos^2 a - \cos^2 b - \cos^2 c + 2 \cos a \cos b \cos c\right\}.$$

Taking square roots, then dividing by $\sin a$ gives

$$\frac{\sin A}{\sin a} = \frac{1}{\sin a \sin b \sin c}\sqrt{1 - (\cos^2 a + \cos^2 b + \cos^2 c) + 2 \cos a \cos b \cos c}.$$

Symmetry in the RHS shows that this value is also produced by the ratios $\frac{\sin B}{\sin b}, \frac{\sin C}{\sin c}$. $\qquad\square$

A.5   Areas of lunes and triangles

Lunes

Refer to Fig. A.5.1.

A lune is the area intercepted on a sphere by two planes meeting on a common diameter. A lune $AXA'YA$ is shown: Angle A is the angle between the planes AXA', AYA' through O. By simple proportion,

$$\frac{\text{lune area } AXA'YA}{4\pi R^2} = \frac{A}{2\pi}, \text{ since } 4\pi R^2 \text{ is the}$$
area of the sphere.

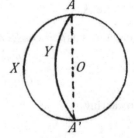

Fig. A.5.1

$$\therefore \text{lune area} = 2AR^2.$$

Triangles

Refer to Fig. A.5.2.

The area of a spherical triangle ABC is defined to be the area Δ of the common region (intersection) of the three lunes L_1, L_2, L_3, where $L_1 = ABUA'CA,\ldots.$

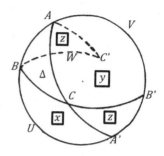

Fig. A.5.2

Consider spherical triangle ABC with area $= \Delta$. Let A', B', C' be diametrically opposite to A, B, C respectively. Δ is the common region (i.e. intersection) of three lunes, viz.,

$L_1 \equiv ABUACA, L_2 \equiv BCBVAB, L_3 \equiv CACWBC.$

L_1 consists of Δ and the part $BUA'CB$, marked x,

$$\text{area} L_1 = 2AR^2 \text{ (see above)}$$
$$= \Delta + x; \tag{A.2}$$

L_2 consists of Δ and the part $CB'VAC$, marked y,

$$\text{area} L_2 = 2BR^2$$
$$= \Delta + y; \tag{A.3}$$

L_3 consists of Δ and the part $C'WBAC'$ (shown dotted), marked z. It is clear that $A'B'C$ has the same area, z, as ABC'.

$$\text{area} L_3 = 2CR^2$$
$$= \Delta + z. \tag{A.4}$$

Theorem [Girard, 1629; Cavalieri, 1632]

Theorem A.5 (Girard–Cavalieri Theorem). Area $ABC = \Delta = (A + B + C - \pi)R^2$.

Proof. Adding (A.2), (A.3), (A.4) gives

$$2\Delta + (\Delta + x + y + z) = 2(A + B + C)R^2.$$

But $\Delta + (x + y + z)$ = area of (the visible front) half of the sphere

$$= 2\pi R^2.$$

$\therefore \Delta + \pi R^2 = (A + B + C)R^2$, whence answer.   □

It is easy to give simple examples of triangles where $A + B + C$ exceeds π. $A + B + C - \pi$ is termed the 'spherical excess'.

A.6   Problems / exercises

1) How long would it take to fly between Bogotà, Colombia ($4°38'N$, $74°05'W$) and Madrid, Spain ($40°25'N$, $3°43'W$) in a plane travelling with an average speed of $800\,\mathrm{kmh}^{-1}$ at a height of $10\,\mathrm{km}$ above sea-level? [$R = 6378\,\mathrm{km}$]

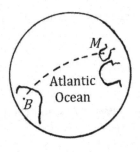

Solution. Refer to Fig. A.6.1.

The latitudes are $\theta = 4°38'$, $\theta' = 40°25'$. Corresponding longitudes are $\varphi = -74°05'$, $\varphi = -3°43'$.

Fig. A.6.1

By the formula (A.1) at the end of A.2, geodesic distance from B to M is given by

$$(6378 + 10)\cos^{-1}[\cos 4°38' \cos 40°25' \cos(74°05' - 3°43')$$
$$+ \sin 4°38' \sin 40°25']$$

$$= 6388 \cos^{-1}[0.3073]$$

$$= 6388 \times 1.258$$

$$\simeq 8039\,\mathrm{km}.$$

Time taken by plane $\simeq \frac{8039}{800}\,\mathrm{h} = 10.05\,\mathrm{h}$; about 10 hours.

Exercise. Compute these distances between cities:

(i) London $(51°31'N, 00°07'E)$, New York $(40°45'N, 73°57'W)$;

(ii) Kuala Lumpur $(3°08'N, 101°42'E)$, Melbourne $(37°51'S, 144°56'E)$.

<u>Ans:</u> (i) 5590 km; (ii) 6370 km. (3 s.f.)   ◇

2) Two ports are on the same parallel of latitude, θ, and the difference in their longitudes is 2φ. Show that the saving of distance in sailing from one to the other on a great circle instead of sailing E.–W. (along a latitude line) is

$$2R\left\{\varphi\cos\theta - \frac{1}{2}\cos^{-1}(1 - 2\sin^2\varphi\cos^2\theta)\right\}.$$

Solution. Refer to Fig. A.6.2.

P_1, P_2 are the ports. P_1P_2 indicates the route along latitude θ. The dotted line gives the shortest route, over a great circle. The distance along the parallel of latitude

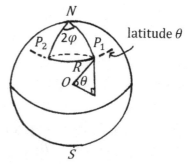

latitude θ

Fig. A.6.2

$$\theta = \left(\tfrac{2\varphi}{2\pi}\right)2\pi R\cos\theta = 2R\varphi\cos\theta.$$

By the formula (A.1) in A.2, the shortest distance (along a great circle) is given by

$$R\cos^{-1}\left\{\cos^2\theta\cos 2\varphi + \sin^2\theta\right\}$$
$$= R\cos^{-1}\left\{\cos^2\theta(1 - 2\sin^2\varphi) + \sin^2\theta\right\}$$
$$= R\cos^{-1}\left\{1 - 2\cos^2\theta\sin^2\varphi\right\},$$
∴ distance saved $= 2R\left\{\varphi\cos\theta - \frac{1}{2}\cos^{-1}(1 - 2\sin^2\varphi\cos^2\theta)\right\}.$

Exercise. Show the above result is more simply written as
$$2R\left\{\varphi\cos\theta - \sin^{-1}(\cos\theta\sin\varphi)\right\}.$$

◇

3) A ship starts from a point on the equator and sails in a great circle cutting the equator at an angle of $45°$. By how much has its longitude changed when it has reached a latitude of $\tan^{-1}\left(\tfrac{1}{2}\right)$?

Solution. Refer to Fig. A.6.3.

PQ is the path taken by the ship. NQK is the meridian through Q cutting the equator at the point K.

$Q\hat{P}K = 45°, QK = \tan^{-1}\left(\frac{1}{2}\right)$. For $P, \theta = 0$; for $Q, \theta' = \tan^{-1}\left(\frac{1}{2}\right)$, and $L = \varphi - \varphi'$ is the longitude change required.

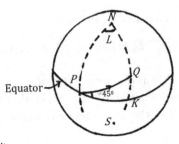

Fig. A.6.3

$$PQ = \cos^{-1}(\cos\theta\cos\theta'\cos L + \sin\theta\sin\theta')$$
$$= \cos^{-1}\left(\frac{2}{\sqrt{5}}\cos L\right).$$

Using the sine rule on QPK,

$$\frac{\sin PQ}{\sin 90°} = \frac{\sin QK}{\sin 45°},$$

$$\therefore \sin PQ = \frac{1/\sqrt{5}}{1/\sqrt{2}} = \frac{\sqrt{2}}{\sqrt{5}}, \text{ whence } \cos PQ = \frac{\sqrt{3}}{\sqrt{5}} = \frac{2}{\sqrt{5}}\cos L,$$

$$\therefore \cos L = \frac{\sqrt{3}}{2}, \text{ so } L = 30°.$$

4) On a sphere where $R = 1$, a triangle is taken with $a = 0.95$, $b = 1.25$, $C = 1.65$. Find c and A, B.

Solution. $\cos c = \cos C \sin a \sin b + \cos a \cos b$
$$= \cos(1.65)\sin(0.95)\sin(1.25) + \cos(0.95)\cos(1.25)$$
$$\simeq 0.1223.$$

Thus $c \simeq 1.448$.

$$\cos A = \frac{\cos a - \cos b \cos c}{\sin b \sin c}$$
$$= \frac{\cos(0.95) - \cos(1.25) \times 0.1223}{\sin(1.25)\sin(1.448)}, \text{ using values found,}$$
$$= 0.5767.$$

Thus $A \simeq 0.956$ ($\simeq 54.8°$).

$$\cos B = \frac{\cos b - \cos a \cos c}{\sin a \sin c}$$

$$= \frac{\cos(1.25) - \cos(0.95) \times 0.1223}{\sin(0.95)\sin(1.448)}$$

$$= 0.3025.$$

Thus $B \simeq 1.264$ ($\simeq 72.4°$).

Exercise.

(i) After finding c here by the cosine rule, use the sine rule in A.4 to show that $\sin A \simeq 0.8170$ and $\sin B \simeq 0.9532$ and thus confirm the values of A, B in the solution above.

(ii) The Bermuda Triangle is a sea area bounded roughly by Bermuda ($33°N$, $65°W$), Miami ($26°N$, $80°W$) and Puerto Rico ($18°N$, $67°W$). By first finding a, b, c for this triangle from A.2, and finding the angles A, B, C from the cosine rule in A.3, followed by the Girard–Cavalieri theorem in A.5 (Theorem A.5), calculate the sea area covered by the triangle.

<u>Ans:</u> About 1.18 million km^2.   ◇

Author Index

Book References

Below are some of the books which may be consulted.

Advanced texts

(1) E. W. Hobson
"*A Treatise on Plane and Advanced Trigonometry*" (7th Edition, 1928)
Dover Publications, Inc. New York; 1957 Reprint

(2) I. Todhunter & J. G. Leathem
"*Spherical Trigonometry*" (1st Edition, 1901)
Macmillan & Co. Ltd., London; 1949 Reprint

School/College & Intermediate texts

(3) C. V. Durell & A. Robson
"*Advanced Trigonometry*" (1st Edition, 1930)
G. Bell & Sons Ltd., London; 1961 Reprint

(4) H. S. Hall & F. H. Stevens
"*A School Geometry, Parts I-IV*" (2nd Edition, 1904)
Macmillan & Co. Ltd., London; 1965 Reprint

(5) H. S. Hall & S. R. Knight
"*Elementary Trigonometry*" (4th Edition, 1905)
Macmillan & Co. Ltd., London; 1955 Reprint

(6) S. L. Loney
    "Plane Trigonometry Vol. I" (6th Metric Edition, 2008) &
    "Analytical Trigonometry Vol. II"
    Maxford Books, New Delhi; 2010 Reprint

(7) F. G. Brown
    "Plane Trigonometry" (1st Edition, 1930)
    Macmillan & Co. Ltd., London

(8) H. S. M. Coxeter
    "Introduction to Geometry" (2nd Edition, 1980)
    John Wiley & Sons, Inc.

Appendix B

More Exercises for Practice

Full solutions to the following are given immediately after the end of this list.

(1) Point P is at an unknown distance east of O, and Q is south of P, a distance of $55\,\mathrm{m}$. OQ is known to be $73\,\mathrm{m}$. Find OP.

(2) $\triangle ABC$ is right-angled at C, and p is the perpendicular length from C to AB. Show, by similar triangles or otherwise that $\frac{p}{b} = \frac{a}{c}$. Deduce that $\frac{1}{p^2} = \frac{1}{a^2} + \frac{1}{b^2}$.

(3) Two circles with radii r, R $(r < R)$, and centres C, D are in contact externally. PQ is their external common tangent.

Show that $PQ^2 = (R+r)^2 - (R-r)^2$, and hence find PQ.

When $R = 3r$, explain why $C\hat{D}Q = 60°$. Calculate exactly the area bounded by PQ and the circles.

(4) If θ is an obtuse angle, give as an exact fraction $\cos\theta$, when $\sin\theta = \frac{45}{53}$. Give θ in degrees, to 1 d.p.

(5) From the pair of equations $\cos\theta + \cos\varphi = a$, $\sin\theta + \sin\varphi = b$, prove that $a\cos\theta + b\sin\theta = \frac{1}{2}(a^2 + b^2)$.

(6) The sides of $\triangle ABC$ are: $a = 69, b = 139, c = 160$. Show, from the cosine rule, that one angle is $60°$ and find the other angles.

147

(7) Quote a formula for median length AA' of a $\triangle ABC$. $ABCD$ is a parallelogram with $AB = 79\,\text{cm}$, $BC = 47\,\text{cm}$, $CA = 104\,\text{cm}$. Calculate BD and the acute angle between the diagonals.

(8) $\triangle ABC$ has $a = 6$, $B = 52°$, $C = 44°$. Find the radius of the circumcircle and the area of $\triangle ABC$.

(9) Give Hero's formula for the area of $\triangle ABC$.

A triangle has $a = 5, b = 5, c = 2$. A second triangle $A'B'C'$ has the same perimeter and area as the first.

If $B'C' = 3$ find the other sides $C'A'$ and $A'B'$.

(10) (a) Prove that the area of $\triangle ABC = 2R^2 \sin A \sin B \sin C$, where R is the circumradius;

(b) Perpendiculars from A, B, C to the opposite sides of $\triangle ABC$ meet the circumcircle in D, E, F. Find the angles of $\triangle DEF$ and prove

$$\frac{\text{area}\,\triangle DEF}{\text{area}\,\triangle ABC} = 8 \cos A \cos B \cos C.$$

(11) A helicopter travelling at a constant speed follows an upward path at an angle θ to the horizon. It passes over point A immediately over an observation point O. From O, it is seen that a minute later it is at point P at elevation α, and after another minute it is at point Q at elevation β. Show that θ, α, β are related by $\tan \theta = 2 \tan \beta - \tan \alpha$.

(12) Given $\sin 54° = \frac{1}{4}(\sqrt{5} + 1)$, with the aid of addition formulae, prove that: $4(\cos 66° + \sin 84°) = \sqrt{15} + \sqrt{3}$.

(13) The elevation of the top, T, of a skyscraper OT at a point W, west of the building, is θ; and at a point N, north of the building, is φ. From a point P halfway along WN the elevation is α.

Show that $\tan \alpha = 2 \tan \theta \tan \varphi / \sqrt{\tan^2 \theta + \tan^2 \varphi}$.

If $\theta = 30°$ and $\varphi = 60°$, find α.

(14) Show that $\cot 15° = 2 + \sqrt{3}$ and $\operatorname{cosec} 15° = \sqrt{6} + \sqrt{2}$.

Deduce that $\cot 7\frac{1}{2}° = 2 + \sqrt{3} + \sqrt{6} + \sqrt{2}$.

Find a similar formula for $\cot 37\frac{1}{2}°$.

(15) Simplify $\cos^2 \theta + \cos^2(60° + \theta) + \cos^2(60° - \theta)$.

An equilateral $\triangle ABC$ has centre O. The radius of the circumcircle is R. A point P is taken on the minor arc BC and PA, PB and PC are drawn. If $O\hat{A}P = \theta$, find, in terms of θ angles $O\hat{P}B$, $O\hat{P}C$. Hence, give the lengths PA, PB, PC.

Show that $PA^2 + PB^2 + PC^2 = 6R^2$.

(16) Use Napier's rule to solve $\triangle ABC$ when $A = 62°, b = 37, c = 18$.

(17) Express $\tan 2A$ in terms of $\tan A$.

Prove that $\tan 70° = \tan 20° + 2\tan 40° + 4\tan 10°$.

(18) Write down expressions for $\sin \theta$, $\cos \theta$, $\tan \theta$ in terms of $t = \tan \frac{\theta}{2}$.

If $\tan^3 \frac{\varphi}{2} = \tan \frac{\theta}{2}$, show that $\tan \varphi = 2\tan \frac{\theta+\varphi}{2}$.

(19) (a) Prove that:
$$\cos 2\theta + \cos 2\varphi - \cos(\theta + \varphi) = 4\cos(\theta + \varphi) \times$$
$$\sin\left(30° - \frac{\theta - \varphi}{2}\right)\sin\left(30° + \frac{\theta - \varphi}{2}\right).$$
(b) If $A + B + C = 180°$, prove that:
$$\cos 2A + \cos 2B + \cos 2C = -1 - 4\cos A \cos B \cos C.$$

(20) Triangle ABC has sides $a = 5$, $b = 7$, $c = 8$. Use a half-angle formula to determine the angle opposite the longest side.

(21) Solve the following equations for θ, in the range from $0°$ to $180°$:

  (a) $\sin\theta + \cos\theta = \sqrt{\frac{3}{2}}$;

  (b) $\tan\theta + \cot\theta = 2\sqrt{2}$;

  (c) $7\sin\theta - 24\cos\theta = 20$.

(22) (a) Prove that $\tan^{-1}\left(\frac{1}{3}\right) - \tan^{-1}\left(\frac{1}{5}\right) = \tan^{-1}\left(\frac{3}{11}\right) - \tan^{-1}\left(\frac{1}{7}\right)$;

    (b) Solve the equation: $\tan^{-1}\left(\frac{2x+1}{3x+3}\right) + \tan^{-1}\left(\frac{3x+2}{5x+3}\right) = \frac{\pi}{4}$.

(23) A quadrilateral $ABCD$ is inscribed in a circle with diameter d. If $AB = AD$, $CB = CD$ and if BD intersects AC at K where $AK = \frac{1}{3}AC$, prove

  (i) the area of $ABCD = \frac{1}{3}d^2\sqrt{2}$;

  (ii) the radius of the inscribed circle $= \frac{(2-\sqrt{2})d}{\sqrt{3}}$.

(24) A person on a cliff $40\,\text{m}$ high can just notice the top of a ship's mast $10\,\text{m}$ high. Find the distance between ship and cliff. (The Earth's radius, $R \cong 6400\,\text{km}$.)

(25) Express the product of the diametral lengths of the circumcircle and incircle of $\triangle ABC$ in terms of a, b, c.

(26) In $\triangle ABC$, prove $r_b + r_c = a\cot\dfrac{A}{2}$.

(27) In $\triangle ABC$, prove

  (i) $\Delta = \dfrac{abc}{s}\cos\dfrac{A}{2}\cos\dfrac{B}{2}\cos\dfrac{C}{2}$;

  (ii) $rr_ar_br_c = \Delta^2$.

(28) In $\triangle ABC$, $\left(1 - \dfrac{r_a}{r_b}\right)\left(1 - \dfrac{r_a}{r_c}\right) = 2$. Show that $A = 90°$.

(29) In $\triangle ABC$, prove that $r_a + r_b + r_c - r = 4R$.

(30) In $\triangle ABC$, prove that $\dfrac{ab - r_ar_b}{r_c} = 4R\sin\dfrac{A}{2}\sin\dfrac{B}{2}\sin\dfrac{C}{2}$.

(31) Find the direct distance between:

    (a) Singapore ($01°17'N$, $103°48'E$) and Melbourne, Australia ($37°51'S$, $144°56'E$);

    (b) Paris, France ($48°52'N$, $02°19'E$) and New York ($40°45'N$, $73°57'W$).

    Take the radius of the Earth as 6370 km.

(32) A, P, Q are three positions on the Earth's surface. A and Q are on the equator, Q being $45°E$ of A; and P is on latitude $60°N$, exactly N. of Q. Find the distance AP, given that the Earth's radius is 6370 km.

(33) ABC is a spherical triangle on a sphere of radius 1. With the usual notation, it has sides $a = 0.8$, $b = 1.4$; and $C = 1.5$ rads. Use the cosine rule to find side c, and the sine rule to find both A and B.

(34) From the equations $x \sec \theta = 2a \sin \theta$, $(b + y) \tan^2 \theta = b - y$, eliminate θ to find a relation connecting x, y with a, b.

(35) Solve the pair of equations $\begin{cases} \cos \theta +5\cos \varphi = 5, \\ \sin \theta \qquad\; = 2\sin \varphi, \end{cases}$
for angles θ, φ. (both less than $180°$)

(36) A tower OT stands on level ground. At a distance x from the base O is the foot F of a hill sloping upwards away at angle α to the horizontal. A person P on the hill overlooks the top T of the tower and notices a pond Q distant y from the tower. If $PF = z$, show that
$$OT = \frac{yz \sin \alpha}{x + y + z \cos \alpha}.$$

(37) A vertical wall $ABCD$, of rectangular shape, and of length $AB = d$ is seen from a position O opposite the bottom end A on the ground. The angles of elevation of D and C are θ, φ respectively. Find the distance OA.

(38) A man walks north up the line of greatest slope of an inclined plane, from A to B, a distance of 100 m. He then walks north-east for 80 m

to C. If the angle of the plane is $20°$, find the inclination of BC to the horizontal, and the height of C above the level of A.

(39) The angular elevation of an aeroplane is observed at times $t = 0$, $t = T$, $t = 2T$ to have values α, β, γ. The aeroplane is flying at constant speed v relative to the ground, and at a height H.
Show that $v = \frac{H}{T}\left(\frac{1}{2}\cot^2\alpha - \cot^2\beta + \frac{1}{2}\cot^2\gamma\right)^{\frac{1}{2}}$.

(40) A lighthouse, height L metres, stands at the edge of a vertical cliff h metres high. From a boat B the angle of elevation of the top T of the lighthouse is α. Another boat C, x metres from B, lies between B and the foot O of the cliff. From C, the angles of elevation of the foot F and top T of the lighthouse are α, β.
Prove that

$$L\tan\alpha = h(\tan\beta - \tan\alpha) \text{ and } h = \frac{x\tan^2\beta}{\tan\beta - \tan\alpha}.$$

Given $h = 100$, $\alpha = 32°$, $\beta = 46°$, find L and h.

(41) A straight path rises at an angle θ to the horizontal. O, P and Q are three points on the path, P being higher than O and Q being higher than P. The distance $OP = x$ metres. At Q there is a vertical pole QR, the height of R above $Q = h$ metres. Prove that, if QR subtends angles α, β at O, P respectively, then

$$h = \frac{x\sin\alpha\sin\beta}{\sin(\beta - \alpha)\cos\theta}.$$

Prove also that the height of R above $O = \dfrac{x\sin\beta\sin(\alpha + \theta)}{\sin(\beta - \alpha)}$.

(42) A tower 51 metres high has a mark at a height of 25 metres from the ground. To an eye 5 metres from the ground x metres from the base, equal angles are subtended by the two parts. Show that

$$2\tan^{-1}\left(\frac{20}{x}\right) = \tan^{-1}\left(\frac{46}{x}\right) - \tan^{-1}\left(\frac{5}{x}\right).$$

Solve the equation to find x.

(43) From a point P on the sloping face of a hill, two straight paths PH, PK are drawn, one in a vertical plane due South, the other in a vertical plane due East. The angle between the paths $H\hat{P}K$ is α,

and the lengths to the horizontal road HK at the foot of the hill are $PH = a, PK = b$. If the slope of the hill $= \theta$, show that

$$\sin^2 \theta = \frac{a^2 + b^2 - 2ab \cos \alpha}{ab \sin \alpha \tan \alpha}.$$

(44) In a convex quadrilateral $OACB$, the angles A and B are right angles, the angle at O is θ and $OA = a$, $OB = b$. The lines AB, OC cut at X. Prove:

(i) $AC \sin \theta = b - a \cos \theta$;

(ii) $OC \sin \theta = \sqrt{a^2 + b^2 - 2ab \cos \theta}$;

(iii) $\dfrac{XC}{OX} = \dfrac{(a - b \cos \theta)(b - a \cos \theta)}{ab \sin^2 \theta}$.

(45) One side of a hill has the form of an inclined plane, the line of greatest slope up, AB being due north. A man starts from A and walks to B, a distance d_1, metres. At B he walks d_2 metres east in a straight line inclined at an acute angle θ to AB, reaching a point C. The heights of B and C above the level of A are h_1, h_2. Prove:

$$\cos \theta = \frac{d_1(h_2 - h_1)}{h_1 d_2}.$$

The bearing of C from A is east of north. Prove

$$\cos \varphi = \frac{h_1 d_2 \sin \theta}{h_2 \sqrt{d_1^2 - h_1^2}}.$$

(46) (a) Prove $2 \tan^{-1} \left(\frac{8}{15} \right) = \sin^{-1} \left(\frac{240}{289} \right)$;

(b) Solve $\cot^{-1} x + \cot^{-1}(2x) = \frac{3\pi}{4}$. $(x > 0)$

(47) (a) Eliminate θ between the equations

$$\operatorname{cosec} \theta - \sin \theta = a^3, \quad \sec \theta - \cos \theta = b^3.$$

(b) Show that if $\sin \theta + \cos \theta = a$, $\sin 2\theta + \cos 2\theta = b$, then

$$(a^2 - b - 1)^2 = a^2(2 - a^2).$$

(48) If $\tan 2\theta + \tan 2\varphi = 0$, show that $\tan \theta = \cot \varphi$.

If, further, $\tan \theta + \tan \varphi = 2.5$, find θ and φ as acute angles.

(49) Solve for θ: $\tan\theta - \cot 2\varphi = \tan\varphi - \cot 2\theta = 1\frac{1}{4}$.

Hint: put $x = \tan\theta$, $y = \tan\varphi$ and solve the resulting equations.

(50) In the usual notation for $\triangle ABC$, show that each of the following expressions

$$\sqrt{rr_ar_br_c}, \quad rr_a \cot\frac{A}{2}, \quad Rr(\sin A + \sin B + \sin C),$$
$$\cos\frac{A}{2}\sqrt{bc(s-b)(s-c)}, \quad s(s-a)\tan\frac{A}{2}.$$

gives the area of $\triangle ABC$.

(51) (a) The area of a triangle is 96, and the radii of the escribed circles are 8, 12, 24. Find the sides.

(b) In a triangle ABC, $a = 13, b = 14, c = 15$. Find r and R.

(52) A yacht travels across the Arabian Sea from Aden, Yemen ($12°51'N$, $45°05'E$) seeking to reach Mumbai, India ($18°56'N$, $72°51'E$). Its average speed on a direct route there is 15 km/h. Leaving Monday morning, when was it expected to arrive?

(Earth's radius $\simeq 6380\,\text{km}$).

Appendix C

Solutions to the More Exercises for Practice

(1) Since $O\hat{P}Q = 90°, OP^2 = 73^2 - 55^2 = 128 \times 18 = 8^2 \times 6^2$, so $OP = 48$ cm.

(2) $\frac{1}{2}p \cdot AB = \Delta = \frac{1}{2}BC \cdot AC, \therefore pc = ab$, and

$$\frac{1}{p^2} = \frac{c^2}{a^2b^2} = \frac{a^2 + b^2}{a^2b^2} = \frac{1}{a^2} + \frac{1}{b^2}.$$

(3) Draw CX perpendicular to DQ. $CXQP$ is a rectangle. $CD = R + r$, $XD = R - r$, so

$$PQ^2 = CX^2 = CD^2 - XD^2 = (R + r)^2 - (R - r)^2.$$

Hence $PQ = \sqrt{4Rr}$.

When $R = 3r$, $PQ = 2r\sqrt{3}$. Since $XD = \frac{1}{2}CD, C\hat{D}X = 60°$.

Area trapezium $CDQP = \frac{1}{2}(r + 3r)2r\sqrt{3} = 4r^2\sqrt{3}$.

The sectors subtracted off have total area $\frac{1}{2}r^2\frac{2\pi}{3} + \frac{1}{2}(3r)^2\frac{\pi}{3} = \frac{11\pi r^2}{6}$.

$\therefore$ Required area$= r^2\left(4\sqrt{3} - \frac{11\pi}{6}\right)$.

(4) $\cos\theta = -\sqrt{1 - \left(\frac{45}{53}\right)^2} = -\frac{1}{53}\sqrt{53^2 - 45^2} = -\frac{1}{53}\sqrt{98 \times 8} = -\frac{28}{53}$,

$\theta = 121.9°$.

(5) Multiply the 1st equation by $\cos\theta$, the 2nd equation by $\sin\theta$ and add:

$$a\cos\theta + b\sin\theta = \cos\theta(\cos\theta + \cos\varphi) + \sin\theta(\sin\theta + \sin\varphi)$$
$$= 1 + (\cos\theta\cos\varphi + \sin\theta\sin\varphi).$$

Also, $a^2 + b^2 = (\cos\theta + \cos\varphi)^2 + (\sin\theta + \sin\varphi)^2$
$$= (\cos^2\theta + \sin^2\theta) + (\cos^2\varphi + \sin^2\varphi) + 2(\cos\theta\cos\varphi + \sin\theta\sin\varphi)$$
$$= 2[1 + (\cos\theta\cos\varphi + \sin\theta\sin\varphi)].$$

Hence result.

(6) $\cos B = \dfrac{160^2 + 69^2 - 139^2}{2 \times 160 \times 69} = \dfrac{11040}{22080} = \dfrac{1}{2}$, so $B = 60°$.

Again $\cos C = \dfrac{69^2 + 139^2 - 160^2}{2 \times 69 \times 139} = -\dfrac{1518}{19182} \simeq -0.0791$,

so $C = 94.5°$, $A = 180° - (60° + 94.5°) = 25.5°$.

(7) $AA' = \dfrac{1}{2}(2b^2 + 2c^2 - a^2)^{\frac{1}{2}}$.

In the parallelogram, $BD = (2 \times 79^2 + 2 \times 47^2 - 104^2)^{\frac{1}{2}}$
$$= \sqrt{6084}$$
$$= 78\,\text{cm},$$
$\cos\theta = (78^2 + 52^2 - 47^2)/(2 \times 78 \times 52) \simeq 0.8110; \theta \simeq 35.8°$.

(8) The 3rd angle $= 84°$. By the sine rule,

$$\frac{6}{\sin 84°} = 2R = \frac{c}{\sin 44°},$$

$R = 3.017$ and $c = 4.191$, $\Delta = \frac{1}{2} \times 6 \times 4.191 \times \sin 52° \simeq 9.91$.

(9) Bookwork: $\Delta = \sqrt{s(s-a)(s-b)(s-c)}$ in usual notation.

Perimeter $= 5 + 5 + 2 = 12$; $s = 6$,

$\Delta^2 = 6 \times 1 \times 1 \times 4 = 24$. The 2nd Δ has $s = 6$ and $\Delta^2 = 24$.

For 2nd Δ, $a = 3$, $\therefore b + c = 9$ and $\Delta^2 = 6 \times 3 \times (6 - b)(6 - c) = 24$.

$\frac{4}{3} = 36 - 6(b + c) + bc = -18 + bc, bc = 19\frac{1}{3}$.

$(b-c)^2 = (b+c)^2 - 4bc = 81 - 77\frac{1}{3} = 3\frac{2}{3}.$

$\therefore b - c \simeq 1.915.$ Solving with $b + c = 9$ gives $b \simeq 5.46, c \simeq 3.57.$

(10) (i) $\Delta = \frac{1}{2}ab \sin C = \frac{1}{2}(2R \sin A)(2R \sin B) \sin C$
$= 4R^2 \sin A \sin B \sin C.$

(ii) $A\hat{D}E = A\hat{B}E = 90° - A$, $A\hat{D}F = A\hat{C}F = 90° - A$ also;
$\therefore E\hat{D}F = 180° - 2A.$

Thus $\hat{D} = 180° - 2A$; $\hat{E} = 180° - 2B$, $\hat{F} = 180° - 2C$ also;

$\triangle DEF$ has the same circumradius as $\triangle ABC$, viz. R, so using (i),

area $\triangle DEF = 4R^2 \sin(180° - 2A) \sin(180° - 2B) \sin(180° - 2C)$
$= 4R^2 \sin 2A \sin 2B \sin 2C$
$= 4R^2 (2\sin A \cos A)(2 \sin B \cos B)(2 \sin C \cos C),$

so ratio of areas $\dfrac{\text{area} \triangle DEF}{\text{area} \triangle ABC} = 8 \cos A \cos B \cos C.$

(11) From $\triangle OAP$,

$$\frac{d}{\sin(90° - \alpha)} = \frac{OP}{\sin(90° + \theta)},$$

where $d = AP = PQ$, since the speed is constant. From $\triangle OAQ$, similarly find

$$\frac{2d}{\sin(90° - \beta)} = \frac{OQ}{\sin(90° + \theta)}.$$

Thus

$$OP = \frac{d \cos \theta}{\cos \alpha} \text{ and } OQ = \frac{2d \cos \theta}{\cos \beta}.$$

Now $d \sin \theta = OQ \sin \beta - OP \sin \alpha,$
$\therefore \sin \theta = \dfrac{2 \cos \theta}{\cos \beta} \sin \beta - \dfrac{\cos \theta}{\cos \alpha} \sin \alpha,$ giving $\tan \theta = 2 \tan \beta - \tan \alpha.$

(12) $4(\cos 66° + \sin 84°) = 4[\cos(120° - 54°) + \sin(30° + 54°)]$

$$= 4\left[-\frac{1}{2}\cos 54° + \frac{\sqrt{3}}{2}\sin 54° + \frac{1}{2}\cos 54° \right.$$

$$\left. + \frac{\sqrt{3}}{2}\sin 54° \right]$$

$$= 4\sqrt{3}\sin 54°$$

$$= \sqrt{3}(\sqrt{5} + 1)$$

$$= \sqrt{15} + \sqrt{3}.$$

(13) $OW = OT\cot\theta, ON = OT\cot\varphi$, so

$$OP = \frac{1}{2}OT\sqrt{\cot^2\theta + \cot^2\varphi},$$

$$\tan\alpha = \frac{OT}{OP} = \frac{2}{\sqrt{\cot^2\theta + \cot^2\varphi}} = \frac{2\tan\theta\tan\varphi}{\sqrt{\tan^2\theta + \tan^2\varphi}}$$

as required. With $\theta = 30°, \varphi = 60°$,

$$\tan\alpha = 2 \cdot \frac{1}{\sqrt{3}} \cdot \sqrt{3} \left/ \sqrt{\frac{1}{3} + 3} \right. = \frac{2}{\sqrt{10/3}} = 1.0954, \alpha = 47.6°.$$

(14) $\cot 15° = \dfrac{1}{\tan 15°} = \dfrac{1}{\tan(45° - 30°)}$

$$= \frac{1 + 1/\sqrt{3}}{1 - 1/\sqrt{3}}$$

$$= \frac{\sqrt{3} + 1}{\sqrt{3} - 1}$$

$$= \frac{1}{2}(\sqrt{3} + 1)^2$$

$$= 2 + \sqrt{3},$$

$$\sin 15° = \sin(45° - 30°) = \frac{1}{\sqrt{2}}\frac{\sqrt{3}}{2} - \frac{1}{\sqrt{2}}\frac{1}{2} = \frac{\sqrt{3} - 1}{2\sqrt{2}}.$$

So that $\operatorname{cosec} 15° = \dfrac{2\sqrt{2}}{\sqrt{3} - 1} = \sqrt{2}(\sqrt{3} + 1) = \sqrt{6} + \sqrt{2}.$

Summing,

$$2 + \sqrt{3} + \sqrt{6} + \sqrt{2} = \frac{\cos 15° + 1}{\sin 15°} = \frac{2\cos^2 7\frac{1}{2}°}{2\sin 7\frac{1}{2}° \cos 7\frac{1}{2}°} = \cot 7\frac{1}{2}°.$$

By the same token,

$$\cot 75° = \frac{1}{\cot 15°} = 2 - \sqrt{3} \text{ and since } \sin 75° = \frac{\sqrt{3}+1}{2\sqrt{2}},$$

$$\operatorname{cosec} 75° = \frac{1}{\sin 75°} = \frac{2\sqrt{2}}{\sqrt{3}+1} = \sqrt{2}(\sqrt{3}-1) = \sqrt{6} - \sqrt{2}.$$

Hence $\cot 37\frac{1}{2}^° = \cot 75° + \operatorname{cosec} 75° = 2 - \sqrt{3} + \sqrt{6} - \sqrt{2}.$

(15) $\cos^2 \theta + \cos^2(60° + \theta) + \cos^2(60° - \theta)$

$$= \cos^2 \theta + \left(\frac{\cos\theta}{2} - \frac{\sqrt{3}}{2}\sin\theta\right)^2 + \left(\frac{\cos\theta}{2} + \frac{\sqrt{3}}{2}\sin\theta\right)^2$$

$$= \frac{3}{2}\cos^2\theta + \frac{3}{2}\sin^2\theta$$

$$= \frac{3}{2}.$$

In a figure, $O\hat{P}B = A\hat{P}B + O\hat{P}A = 60° + \theta$, and $O\hat{P}C = A\hat{P}C - O\hat{P}A = 60° - \theta$. This makes $PA = 2R\cos\theta$, $PB = 2R\cos(60° + \theta)$, $PC = 2R\cos(60° - \theta)$. So that $PA^2 + PB^2 + PC^2 = 4R^2(\cos^2\theta + \cos^2(60° + \theta) + \cos^2(60° - \theta)) = 4R^2(3/2) = 6R^2$.

(16) By Napier's rule,

$$\tan\frac{B-C}{2} = \frac{37-18}{37+18} \Big/ \tan\frac{62°}{2} = \frac{19}{55} \Big/ \tan 31° \simeq 0.2076.$$

$$\frac{B-C}{2} \simeq 11.726°, B - C \simeq 23.45°. \text{ But } B + C = 180° - A = 118°.$$

Hence $B \simeq 70.7°$, $C \simeq 47.3°$. Then $a = \sin 62° \times 37/\sin 70.7° \simeq 34.6$.

(17) Bookwork: $\tan 2A = \dfrac{2\tan A}{1 - \tan^2 A}$ $\cdots\cdots\cdots\cdots\cdots\cdots\cdots\cdots\cdots\cdots\cdots\cdots\cdots$ (♣)

Put $\tan 70° = t$. Then $\tan 20° = \frac{1}{t}$. Remains to prove

$$\frac{1}{2}\left(t - \frac{1}{t}\right) = \tan 40° + 2\tan 10°.$$

From (♣), $LHS = \dfrac{t^2 - 1}{2t} = -\dfrac{1}{\tan 140°} = \dfrac{1}{\tan 40°}.$

Put $\tan 40° = T$. Remains to prove $\dfrac{1}{2}\left(\dfrac{1}{T} - T\right) = \tan 10°.$

$$LHS = \frac{1 - T^2}{2T} = \frac{1}{\tan 80°} = \tan 10°.$$

(18) Bookwork (first part).

Next, put $\tan\frac{\varphi}{2} = T, \tan\frac{\theta}{2} = t$, so $T^3 = t$ is given

$$2\tan\frac{\theta+\varphi}{2} = \frac{2(t+T)}{1-tT} = \frac{2(T^3+T)}{1-T^4} = \frac{2T}{1-T^2} = \tan\varphi.$$

(19) (a) By factorisation formulae.

$$2\sin\left(30° - \frac{\theta-\varphi}{2}\right)\sin\left(30° + \frac{\theta-\varphi}{2}\right) = \cos(\theta-\varphi) - \cos 60°.$$

So $RHS = 2\cos(\theta+\varphi)(\cos(\theta-\varphi) - 1/2)$
$= 2\cos(\theta+\varphi)\cos(\theta-\varphi) - \cos(\theta+\varphi)$
$= (\cos 2\theta + \cos 2\varphi) - \cos(\theta+\varphi)$
$= LHS$

(b) Again, by factorisation formulae.

$LHS = \cos 2A + 2\cos(B+C)\cos(B-C)$
$= (2\cos^2 A - 1) - 2\cos A\cos(B-C),$ (since $B+C = 180° - A$)
$= -1 + 2\cos A(\cos A - \cos(B-C))$
$= -1 - 2\cos A(\cos(B+C) - \cos(B-C))$
$= -1 - 4\cos A\cos B\cos C$
$= RHS.$

(20) Semi-perimeter, $s = \frac{1}{2}(5+7+8) = 10$. Angle C is required.
$\cos\frac{C}{2} = \sqrt{\frac{10\times(10-8)}{5\times 7}} = \frac{2}{\sqrt{7}} \simeq 0.7559; C \simeq 2 \times 40.89° \simeq 81.8°.$

(21) (a) Squaring: $1 + 2\sin\theta\cos\theta = \frac{3}{2}, \therefore \sin 2\theta = \frac{1}{2},$
$2\theta = 30°, 150°; \theta = 15°, 75°.$

(b) Put $\tan\theta = t$. Since $t + \frac{1}{t} = 2\sqrt{2}$, so $t^2 - 2\sqrt{2}t + 1 = 0$,
$t = \sqrt{2} \pm 1, \theta = 22\frac{1}{2}°, 67\frac{1}{2}°.$

(c) Divide by $\sqrt{7^2 + 24^2} = 25$. The equation is then $\sin(\theta - \alpha) = 0.8$,
where $\tan\alpha = \frac{24}{7}, \alpha \simeq 73.7°; \theta - \alpha \simeq 53.1°, \theta \simeq 126.8°.$

(22) (a) $\tan(LHS) = \left(\dfrac{1}{3} - \dfrac{1}{5}\right) \Big/ \left(1 + \dfrac{1}{15}\right) = \dfrac{2}{16} = \dfrac{1}{8}$,

$\tan(RHS) = \left(\dfrac{3}{11} - \dfrac{1}{7}\right) \Big/ \left(1 + \dfrac{3}{77}\right) = \dfrac{10}{80} = \dfrac{1}{8}$,

so $LHS = RHS$.

(b) Take tangents on both sides, and eliminating fractions:

$$1 = \frac{(2x+1)(5x+3) + (3x+3)(3x+2)}{(3x+3)(5x+3) - (2x+1)(3x+2)},$$

$$9x^2 + 15x + 7 = 19x^2 + 24x + 9,$$

$$10x^2 - 9x + 2 = 0,$$

$$(5x - 2)(2x - 1) = 0.$$

Thus $x = \frac{2}{5}, \frac{1}{2}$.

(23) AC is a diameter and BD intersects AC at right angles.

$AK = \frac{d}{3}, CK = \frac{2d}{3}$; and $B\hat{A}K = \frac{A}{2}, B\hat{C}K = \frac{C}{2}$.

Since $B = 90°$, $\frac{A}{2} + \frac{C}{2} = 90°$.

Now $AK \tan \frac{A}{2} = BK = CK \tan \frac{C}{2}$, so $\frac{d}{3} \tan \frac{A}{2} = \frac{2d}{3} \tan \frac{C}{2}$, $\tan \frac{A}{2} = \sqrt{2}$, $\tan \frac{C}{2} = \frac{\sqrt{2}}{2}$.

(i) area $ABCD = \dfrac{1}{2} AC \cdot BD = \dfrac{1}{2} d(2BK) = d \cdot BK$

$= d\left(\dfrac{d}{3} \tan \dfrac{A}{2}\right) = \dfrac{1}{3} d^2 \sqrt{2}$.

(ii) The inscribed circle has a centre M (say) which is on diagonal AC. If r =radius of inscribed circle, $d = AC = AM + MC$, $d = \frac{r}{\sin \frac{A}{2}} + \frac{r}{\sin \frac{C}{2}}$.

But $\sin \frac{A}{2} = \frac{\sqrt{2}}{\sqrt{3}}$, $\sin \frac{C}{2} = \frac{\sqrt{2}}{\sqrt{6}} = \frac{1}{\sqrt{3}}$, $\therefore d = r\left(\frac{\sqrt{3}}{\sqrt{2}} + \sqrt{3}\right)$,

$r = \frac{d}{\sqrt{3}} \Big/ (\frac{1}{\sqrt{2}} + 1) = \frac{d\sqrt{2}}{\sqrt{3}} / (1 + \sqrt{2}) = \frac{d}{\sqrt{3}} \sqrt{2}(\sqrt{2} - 1) = \frac{d}{\sqrt{3}}(2 - \sqrt{2})$.

(24) A figure shows that the distance $= R(\theta + \varphi)$, where θ, φ are small angles in radians such that $\cos\theta = \frac{R}{R+40}, \cos\varphi = \frac{R}{R+10}$, in consistent units.

Since R is large, $\cos\theta = \left(1+\frac{40}{R}\right)^{-1} \simeq 1-\frac{40}{R}$ and $\cos\varphi = \left(1+\frac{10}{R}\right)^{-1} \simeq 1-\frac{10}{R}$.

Hence approximately, $1-\frac{\theta^2}{2} = 1-\frac{40}{R}$, $1-\frac{\varphi^2}{2} = 1-\frac{10}{R}$,

$\theta \simeq \sqrt{\frac{80}{R}} = 3.54 \times 10^{-3}$, $\varphi = \sqrt{\frac{20}{R}} = 1.77 \times 10^{-3}$.

Required distance $= 6400 \times (1.77 + 3.54) \times 10^{-3}\,\text{km} \simeq 34\,\text{km}$.

(25) The product $= (2R) \cdot (2r)$.

Now, $r = \frac{\Delta}{s}$, and because $\Delta = \frac{1}{2}ab\sin C = \frac{ab}{2}\frac{c}{2R}$, $4R = \frac{abc}{\Delta}$.

Thus product $= \left(\frac{abc}{2\Delta}\right) \cdot \left(\frac{2\Delta}{s}\right) = \frac{abc}{s} = \frac{2abc}{a+b+c}$.

(26)
$$r_b + r_c = 4R\cos\frac{A}{2}\sin\frac{B}{2}\cos\frac{C}{2} + 4R\cos\frac{A}{2}\cos\frac{B}{2}\sin\frac{C}{2}$$
$$= 4R\cos\frac{A}{2}\left(\sin\frac{B}{2}\cos\frac{C}{2} + \cos\frac{B}{2}\sin\frac{C}{2}\right)$$
$$= 4R\cos\frac{A}{2}\sin\frac{B+C}{2}$$
$$= 4R\cos^2\frac{A}{2}$$
$$= \frac{2a}{\sin A}\cos^2\frac{A}{2}$$
$$= \frac{a}{\sin\frac{A}{2}\cos\frac{A}{2}}\cos^2\frac{A}{2}$$
$$= a\cot\frac{A}{2}.$$

(27) (i) $RHS = \dfrac{abc}{s} \sqrt{\dfrac{s(s-a)}{bc}} \sqrt{\dfrac{s(s-b)}{ca}} \sqrt{\dfrac{s(s-c)}{ab}}$

$\qquad = \dfrac{1}{s} \sqrt{s^3(s-a)(s-b)(s-c)}$

$\qquad = \sqrt{s(s-a)(s-b)(s-c)}$

$\qquad = \Delta$

$\qquad = LHS.$

(ii) $rr_a r_b r_c = \dfrac{\Delta}{s} \dfrac{\Delta}{s-a} \dfrac{\Delta}{s-b} \dfrac{\Delta}{s-c} = \dfrac{\Delta^4}{\Delta^2} = \Delta^2.$

(28) Using $r_a = \dfrac{\Delta}{s-a}, r_b = \dfrac{\Delta}{s-b}, r_c = \dfrac{\Delta}{s-c}$, we are given that

$\left(1 - \dfrac{s-b}{s-a}\right)\left(1 - \dfrac{s-c}{s-a}\right) = 2.$ So,

$$(b-a)(c-a) = 2(s-a)^2,$$

$2(bc - ac - ab + a^2) = (b+c-a)^2 \equiv b^2 + c^2 + 2bc - 2ab - 2ac + a^2,$

reducing to $a^2 = b^2 + c^2$. So $A = 90°$.

(29) $r_a + r_b + r_c - r = \dfrac{\Delta}{s-a} + \dfrac{\Delta}{s-b} + \dfrac{\Delta}{s-c} - \dfrac{\Delta}{s}$

$\qquad = \dfrac{1}{\Delta}\{(s-b)(s-c)s + (s-c)(s-a)s$

$\qquad\qquad + (s-a)(s-b)s - (s-a)(s-b)(s-c)\}$

$\qquad = \dfrac{1}{\Delta}\{(s-b)(s-c)a + s(s-a)(2s - b - c)\}$

$\qquad = \dfrac{a}{\Delta}\{(s-b)(s-c) + s(s-a)\}$

$\qquad = \dfrac{a}{\Delta}\{2s^2 - (a+b+c)s + bc\}$

$\qquad = \dfrac{abc}{\Delta}$

$\qquad = 4R.$

(30) Writing $\sin\frac{A}{2} = X_a, \cos\frac{A}{2} = Y_a; \sin\frac{B}{2} = X_b, \cos\frac{B}{2} = Y_b; \sin\frac{C}{2} = X_c, \cos\frac{C}{2} = Y_c.$

$ab = (2R\sin A)(2R\sin B) = 4R^2(2X_a Y_a)(2X_b Y_b) = 16R^2 X_a X_b Y_a Y_b,$

$r_a r_b = (4R X_a Y_b Y_c)(4R Y_a X_b Y_c) = 16R^2 X_a Y_a X_b Y_b Y_c^2,$

$$\therefore ab - r_a r_b = 16R^2 X_a Y_a X_b Y_b (1 - Y_c^2) = 16R^2 X_a X_b Y_a Y_b X_c^2,$$

$r_c = 4R Y_a Y_b X_c$, so that

$$\frac{ab - r_a r_b}{r_c} = 4R X_a X_b X_c \equiv 4R \sin\frac{A}{2}\sin\frac{B}{2}\sin\frac{C}{2}.$$

(31) (a) The distance Singapore to Melbourne

$$= 6370 \times \cos^{-1}\left[\cos(1.283°)\cos(-37.85°)\cos(103.8° - 144.9°)\right.$$
$$\left. + \sin(1.283°)\sin(-37.85°)\right]$$
$$= 6370 \times \cos^{-1}(0.5805) \simeq 6060\,\text{km}.$$

(b) The distance Paris to N.Y.

$$= 6370 \times \cos^{-1}\left[\cos(48.87°)\cos(40.75°)\cos(2.317° + 73.95°)\right.$$
$$\left. + \sin(48.87°)\sin(40.75°)\right]$$
$$= 6370 \times \cos^{-1}(0.6100) \simeq 5830\,\text{km}.$$

(32) Let O be the Earth's centre. Using the cosine rule, we find

$$\cos A\hat{O}P = \cos 90° \sin 45° \sin 60° + \cos 60° \cos 45° = \frac{\sqrt{2}}{4}.$$

Thus $A\hat{O}P = 1.2094$ rads, so distance

$$AP = 6370 \times 1.2094 \simeq 7700\,\text{km}.$$

(33) $\cos c = \cos C \sin a \sin b + \cos a \cos b$
$$= (\cos 1.5)(\sin 0.8)(\sin 1.4) + (\cos 0.8)(\cos 1.4)$$
$$= 0.1685,$$

$$\therefore c = 1.401.$$

$$\sin A = \sin a \frac{\sin C}{\sin c} = (\sin 0.8)\frac{\sin 1.5}{\sin 1.401} = 0.7260;$$

$$A = 0.8125 = 46.55°;$$

$$\sin B = \sin b \frac{\sin C}{\sin c} = (\sin 1.4)\frac{\sin 1.5}{\sin 1.401} = 0.9973;$$

$$B = 1.497 = 85.80°.$$

(34) From the 1st equation, $\dfrac{x}{a} = 2\sin\theta\cos\theta = \sin 2\theta$.

From the 2nd equation,

$$\frac{y}{b}(\tan^2\theta + 1) = 1 - \tan^2\theta, \quad \frac{y}{b} = \frac{1 - \tan^2\theta}{1 + \tan^2\theta} = \cos 2\theta.$$

Hence $\left(\dfrac{x}{a}\right)^2 + \left(\dfrac{y}{b}\right)^2 = 1$.

(35) Eliminate θ : $(5 - 5\cos\varphi)^2 + (2\sin\varphi)^2 = 1$,

$$25(1 - 2\cos\varphi + \cos^2\varphi) + 4(1 - \cos^2\varphi) = 1,$$

$$21\cos^2\varphi - 50\cos\varphi + 28 = 0,$$

$$\cos\varphi = \frac{25 - \sqrt{625 - 21 \times 28}}{21} \simeq 0.9008,$$

$$\therefore \varphi = 25.73°.$$

Then $\sin\theta = 0.8683, \theta = 60.26°$.

(36) Height of P above ground$= z\sin\alpha$.

Horizontal distance between P and Q $= z\cos\alpha + FO + OQ$
$$= x + y + z\cos\alpha.$$

Their ratio $= \tan O\hat{Q}T = \dfrac{OT}{y}$, whence OT as given.

(37) Height of wall, $AD = OA\tan\theta$,

$\tan\varphi = \frac{AD}{OB}$ where $OB^2 = OA^2 + d^2$,

$$\tan^2\varphi = \frac{OA^2\tan^2\theta}{OA^2 + d^2},$$

$$OA^2(\tan^2\theta - \tan^2\varphi) = d^2\tan^2\varphi,$$

$$\therefore OA = \frac{d\tan\theta}{\sqrt{\tan^2\theta - \tan^2\varphi}}.$$

(38) Let θ = inclination of BC to the horizon.

$$\tan 20° = \frac{\text{height of } C \text{ above } B}{\text{amount } C \text{ is North of } B}$$
$$= \frac{80 \sin \theta}{80 \cos \theta \cos 45°}$$
$$= \sqrt{2} \tan \theta.$$

Thus $\theta = 14.4°$.

Height of C above A = height of C above B + height of B above A
$$= 80 \sin \theta + 100 \sin 20°$$
$$\simeq 54.1 \, \text{m}.$$

(39) Let A, B, C be the projections of the positions of the plane on the ground and O the observer's position. Then $OA = H \cot \alpha$, $OB = H \cot \beta$, $OC = H \cot \gamma$ and $AB = BC = vT$.

By Apollonius' theorem, $OA^2 + OC^2 = 2OB^2 + 2(vT)^2$,

$$H^2(\cot^2 \alpha + \cot^2 \gamma) = 2H^2 \cot^2 \beta + 2(vT)^2,$$

$$\therefore (vT)^2 = H^2\left(\tfrac{1}{2} \cot^2 \alpha - \cot^2 \beta + \tfrac{1}{2} \cot^2 \gamma\right),$$

v follows on taking square roots.

(40) $O\hat{B}T = \alpha = O\hat{C}F$ and $O\hat{C}T = \beta$, $OT \cot \beta = OC = OF \cot \alpha$,

$$\therefore (h + L) \cot \beta = h \cot \alpha,$$

thus $(h + L) \tan \alpha = h \tan \beta$, $L \tan \alpha = h(\tan \beta - \tan \alpha)$.

Also, $x = OB - OC = (L + h) \cot \alpha - h \cot \alpha = L \cot \alpha$,

$$x \tan \alpha = L, \therefore h = \frac{x \tan^2 \alpha}{\tan \beta - \tan \alpha}.$$

Using the given figures,

$$L = \frac{100(\tan 46° - \tan 32°)}{\tan 32°} \simeq 65.7, x = \frac{L}{\tan 32°} \simeq 105.$$

(41) Draw $RK \perp OPQ$. Then $Q\hat{R}K = \theta$, $RK = h\cos\theta$.

Also $O\hat{R}P = \beta - \alpha$. The sine rule applied to $\triangle OPR$ gives

$$\frac{OR}{\sin(180° - \beta)} = \frac{x}{\sin(\beta - \alpha)} \therefore OR = \frac{x\sin\beta}{\sin(\beta - \alpha)}.$$

Since $RK = OR\sin\alpha$, $h\cos\theta = \dfrac{x\sin\alpha\sin\beta}{\sin(\beta - \alpha)}$.

Height of R above $O = OR\sin(\alpha + \theta) = \dfrac{x\sin\beta\sin(\alpha + \theta)}{\sin(\beta - \alpha)}$.

(42) Let OMT be the tower, where O is the base, M is the mark.

Let E be the eye at distance x from OMT.

M is $25 - 5 = 20$ metres above E, and T is $26 + 20 = 46$ metres above E.

Given $O\hat{E}M = M\hat{E}T$,

$$\therefore \tan^{-1}\frac{5}{x} + \tan^{-1}\frac{20}{x} = \tan^{-1}\frac{46}{x} - \tan^{-1}\frac{20}{x} \text{ or}$$

$$2\tan^{-1}\frac{20}{x} = \tan^{-1}\frac{46}{x} - \tan^{-1}\frac{5}{x}.$$

Taking tangents,

$$\frac{2(\frac{20}{x})}{1 - (\frac{20}{x})^2} = \frac{\frac{46}{x} - \frac{5}{x}}{1 + \frac{230}{x^2}}, \text{ reducing to } \frac{40}{x^2 - 400} = \frac{41}{x^2 + 230},$$

$$x^2 = 9200 + 16400 = 25600, \quad x = 160.$$

(43) Take point Q on the road HK, such that the plane OPQ cuts HK at right angles. $\theta = O\hat{Q}P$. Let $OP = h$ and $HK = d$, then

$$OH^2 = a^2 - h^2, \, OK^2 = b^2 - h^2.$$

We have $a^2 + b^2 - 2ab\cos\alpha = d^2 = (a^2 - h^2) + (b^2 - h^2)$,

$$\therefore h^2 = ab\cos\alpha.$$

Also, $\frac{1}{2}PQ \cdot HK = \text{area} \triangle PHK = \frac{1}{2}ab\sin\alpha$,

$$\therefore PQ = \frac{ab\sin\alpha}{d},$$

$$\sin\theta = \frac{h}{PQ},$$

$$\sin^2\theta = \frac{ab\cos\alpha}{\left(\frac{ab\sin\alpha}{d}\right)^2} = \frac{d^2}{ab\sin\alpha\tan\alpha} = \frac{a^2 + b^2 - 2ab\cos\alpha}{ab\sin\alpha\tan\alpha}.$$

(44) $OACB$ is cyclic, OC is a diameter.

(i) Project lengths along OB : $(b - a\cos\theta) = AC\sin(180° - \theta) = AC\sin\theta$;

(ii) $AB = \sqrt{a^2 + b^2 - 2ab\cos\theta}$, and

$$\frac{AB}{\sin\theta} = 2 \times \text{ radius of circle } OACB = OC,$$

$$\therefore OC\sin\theta = \sqrt{a^2 + b^2 - 2ab\cos\theta};$$

(iii) $\dfrac{XC}{OX} = \dfrac{\text{area} \triangle ABC}{\text{area} \triangle ABO} = \dfrac{\frac{1}{2}AC \cdot BC\sin(180° - C)}{\frac{1}{2}AB\sin\theta} = \dfrac{AC \cdot BC}{ab}.$

Noting now, as in (i), that $BC\sin\theta = a - b\cos\theta$,

$$\frac{XC}{OX} = \frac{(b - a\cos\theta)(a - b\cos\theta)}{ab\sin^2\theta}.$$

(45) Let AB meet the horizontal through C at K, and let α be the slope on the hill. $CK = d_2\sin\theta$, $BK = d_2\cos\theta$, and

$$\sin\alpha = \frac{h_1}{d_1} = \frac{\text{height of } C \text{ above } B}{BK} = \frac{h_2 - h_1}{d_2\cos\theta},$$

$$\therefore \cos\theta = \frac{d_1(h_2 - h_1)}{h_1 d_2},$$

$$\tan\varphi = \frac{CK}{AK\cos\alpha},$$

$$AK\cos\alpha = AB\cos\alpha\frac{h_2}{h_1} = \sqrt{d_1^2 - h_1^2}\frac{h_2}{h_1},$$

$$\therefore \tan\varphi = \frac{h_1 d_2\sin\theta}{h_2\sqrt{d_1^2 - h_1^2}}.$$

(46) (a) $\tan(LHS) = \dfrac{2 \times \frac{8}{15}}{1 - (\frac{8}{15})^2} = \dfrac{16 \times 15}{15^2 - 8^2} = \dfrac{240}{161}$,

$\therefore \sin(LHS) = \dfrac{240}{\sqrt{161^2 + 240^2}} = \dfrac{240}{289} = \sin(RHS)$.

(b) Rewrite the equation as

$$\tan^{-1}\frac{1}{x} + \tan^{-1}\frac{1}{2x} = \frac{3\pi}{4}.$$

Take tangents

$$\frac{\frac{1}{x} + \frac{1}{2x}}{1 - \frac{1}{x^2}} = -1,$$

$\therefore 3x = 1 - 2x^2$, and

$$x = \frac{3 + \sqrt{17}}{4}.$$

(47) (a) The two equations give:

$$\frac{1}{\sin\theta} - \sin\theta = a^3, \quad \frac{1}{\cos\theta} - \cos\theta = b^3,$$

so that

$$\frac{\cos^2\theta}{\sin\theta} = a^3, \quad \frac{\sin^2\theta}{\cos\theta} = b^3.$$

$\therefore \dfrac{a^3}{b^3} = \dfrac{\cos^3\theta}{\sin^3\theta}$, so $\frac{a}{b} = \frac{\cos\theta}{\sin\theta}$, and $a^3 b^3 = \sin\theta\cos\theta$.

Thus $a^4 b^2 = \cos^2\theta$ and $a^2 b^4 = \sin^2\theta$.

Hence $1 = a^2 b^2 (a^2 + b^2)$.

(b) From the first equation,

$$a^2 = \sin^2\theta + 2\sin\theta\cos\theta + \cos^2\theta$$
$$= 1 + \sin 2\theta,$$

$\therefore a^2 - 1 - b = -\cos 2\theta$ and $2 - a^2 = 1 - \sin 2\theta$;

$\therefore a^2(2 - a^2) = 1 - \sin^2 2\theta = (a^2 - 1 - b)^2$.

(48) Put $\tan\theta = t$, $\tan\varphi = T$. Then

$$\frac{2t}{1-t^2} + \frac{2T}{1-T^2} = 0,$$

$$t(1-T^2) + T(1-t^2) = 0,$$

$$(t+T)(1-tT) = 0,$$

$\therefore tT = 1$, i.e. $\tan\theta = \cot\theta$.

With $t + T = 2.5$, $t^2 - 2.5t + 1 = 0$, so $t = \frac{1}{2}$ or 2,

$\theta \simeq 26.6°$ or $63.4°$; correspondingly, $\varphi \simeq 63.4°$ or $26.6°$.

(49) Taking the hint, we are given

$$x - \left(\frac{1-y^2}{2y}\right) = y - \left(\frac{1-x^2}{2x}\right) = 1\frac{1}{4}$$

or

$$2xy + y^2 - 1 = \frac{5}{2}y, \quad 2xy + x^2 - 1 = \frac{5}{2}x.$$

Subtracting,

$$y^2 - x^2 = \frac{5}{2}(y - x),$$

$\therefore x + y = \frac{5}{2}$ (assuming $x \neq y$).

Adding instead,

$$x^2 + y^2 + 4xy - 2 = \frac{5}{2}(x + y).$$

Using $x + y = \frac{5}{2}$, this gives $2xy - 2 = 0$, $xy = 1$.

$\therefore x + \frac{1}{x} = \frac{5}{2}$, $x^2 - \frac{5}{2}x + 1 = 0$; $x = 2$ or $\frac{1}{2}$.

$\theta = \tan^{-1} 2$ or $\tan^{-1}\frac{1}{2}$; i.e. $\theta \simeq 26.6°$ or $63.4°$.

Correspondingly, $\varphi \simeq 63.4°$ or $26.6°$.

(50) $\sqrt{rr_ar_br_c} = \sqrt{\dfrac{\Delta}{s}\dfrac{\Delta}{s-a}\dfrac{\Delta}{s-b}\dfrac{\Delta}{s-c}}$

$$= \sqrt{\dfrac{\Delta^4}{\Delta^2}}$$

$$= \Delta.$$

$rr_a \cot \dfrac{A}{2} = \dfrac{\Delta}{s}\dfrac{\Delta}{s-a}\sqrt{\dfrac{s(s-a)}{(s-b)(s-c)}}$

$$= \dfrac{\Delta^2}{\sqrt{s(s-a)(s-b)(s-c)}}$$

$$= \Delta.$$

$Rr(\sin A + \sin B + \sin C) = \dfrac{r}{2}(a+b+c) = rs = \Delta.$

$\cos \dfrac{A}{2}\sqrt{bc(s-b)(s-c)} = \sqrt{\dfrac{s(s-a)}{bc}}\sqrt{bc(s-b)(s-c)}$

$$= \sqrt{s(s-a)(s-b)(s-c)}$$

$$= \Delta.$$

$s(s-a)\tan \dfrac{A}{2} = s(s-a)\sqrt{\dfrac{(s-b)(s-c)}{s(s-a)}}$

$$= \sqrt{s(s-a)(s-b)(s-c)}$$

$$= \Delta.$$

(51) (a) $\Delta = 96$, and $8 = \dfrac{\Delta}{s-a}, 12 = \dfrac{\Delta}{s-b}, 24 = \dfrac{\Delta}{s-c}.$

Hence $s - a = 12, s - b = 8, s - c = 4.$

Adding, $s = 24$, so that $a = 12, b = 16, c = 20.$

(b) $s = \dfrac{1}{2}(13 + 14 + 15) = 21,$

$\Delta = \sqrt{21 \times 8 \times 7 \times 6} = 84.$

$\therefore r = \dfrac{84}{21} = 4$, and

$$R = \dfrac{13 \times 14 \times 15}{4 \times 84} = \dfrac{65}{8} = 8\dfrac{1}{8}.$$

(52) The distance Aden–Mumbai

$$= 6380 \times \cos^{-1} \left[\cos 12.85° \cos 18.93° \cos (72.85° - 45.08°) \right.$$
$$\left. + \sin 12.85° \sin 18.93° \right]$$
$$= 6380 \times \cos^{-1} (0.888) = 3046 \text{ km.}$$

Time of travel$= \frac{3046}{15 \times 24} = 8.46$ days.

The yacht arrives on Tuesday afternoon of the following week.